Nutritional Support after Gastrointestinal Surgery

Donato Francesco Altomare
Maria Teresa Rotelli
Editors

Nutritional Support after Gastrointestinal Surgery

 Springer

Editors
Donato Francesco Altomare
Department of Emergency and Organ
Transplant
University of Bari Aldo Moro
Bari
Italy

Maria Teresa Rotelli
Department of Emergency
and Organ Transplantat
University of Bari Aldo Moro
Bari
Italy

ISBN 978-3-030-16553-6   ISBN 978-3-030-16554-3   (eBook)
https://doi.org/10.1007/978-3-030-16554-3

This Springer imprint is published by the registered company Springer Nature Switzerland AG
The registered company address is: Gewerbestrasse 11, 6330 Cham, Switzerland

Foreword

Surgery on the gastrointestinal tract may directly interfere with its function of alimentation by various means. Removal of intestine or bypass obviously can affect absorption of nutriments and alter water and electrolyte balance. Nutritional assessment and management of the surgical patient are fundamental; they probably reduce complications and certainly promote rapid recovery.

Nutrition may not always be taken seriously enough by surgeons, yet the introduction of parenteral nutrition over forty years ago has saved thousands of lives. Intestinal surgery today is inconceivable without this vital means of support. Maintaining adequate nutrition is an essential component of surgical practice.

Professor Altomare and Dr. Rotelli have devised a book to deal with all aspects of nutrition, which are likely to present in the surgery of the intestine. The book is structured in a format that allows focus on the effects of operations on all levels of the gastrointestinal tract. The anatomy of a given operation is followed by a description of the physiological changes brought about and then followed by a section on management including dietetic advice and intervention. There is a section on the relationship between the hypothalamus and the gastrointestinal tract, the so-called gut-brain axis, and other important physiological aspects of nutrition.

Pathological aspects are considered with full descriptions of the effect of bowel loss in short bowel syndrome of various degrees of severity and of intestinal fistulation, including stomas. The book deals with nutritional risk factors that may affect the outcome of surgery and demonstrates the value of multidisciplinary care particularly in association with the dietitian. The undernourished patient is considered in detail with information on nutritional deficiencies and their management. This involves the strategy of the timing of surgery and the indications for parenteral nutrition. Bariatric operations are now among the most common procedures in intestinal surgery. The influence of obesity and the metabolic consequences of various interventions are considered in detail. The chapters take into account new technical developments in abdominal surgery.

The authorship includes surgeons, nutritionists, and dietitians, all specialists in the field of nutrition. The uniform structure of the book will help the reader to understand the degree of overlap and the differences of nutritional aspects of surgery at various points of the gastrointestinal tract. The book will be of value to all health care professionals working in the field of intestinal surgery. These include surgeons and physicians, dietitians, nurses, and students and trainees in all disciplines. It is an up-to-date textbook, which occupies an important place in the surgical literature.

R. J. Nicholls

St. Mark's Hospital, London, UK

Imperial College, London, UK

Preface

Surgeons and nutritionists are often challenged by the question: *doctor, what can I eat after this surgery?* In this book we sought to answer this foreseeable question. In fact, oral feeding is a major concern of any patient after digestive tract surgery; on the other hand, eating correctly and keeping the body weight under control is essential for the overall patient recovery. Surgery on the digestive tract may affect digestion and absorption physiology: the nutritional approach is a great opportunity to provide patients with a long-lasting educational program, also to correct any erroneous presurgery eating habit. Even after operations with minor physiologic implications, like the proctologic ones, postoperative feeding remains a major concern to improve patient's life quality.

Our aim was ambitious: rather than providing a list of unspecific nutritional prescriptions or general dietetic advice, we tried to explain the changes in the nutritional requirements and dietary approach induced by these operations, looking at the surgery-induced changes in digestive physiology, and to outline the "optimal" nutritional approach taking into account not only the peculiar absorption deficits and the metabolic imbalances generated from the surgical procedure but also the single patient's comorbidities, necessities, and eventually food intolerances.

Nutritionists and dietitians are often not familiar with some types of gastrointestinal operations and therefore in this textbook, before the description of the nutritional requirements, we shortly describe the most frequent types of gastrointestinal surgery and the relative changes induced in the physiology of nutrition.

Surprisingly, the literature on these topics is scanty. Accordingly, we necessarily had to provide some statements based more on the experience of the single experts than on scientific evidences. Artificial enteral and parenteral nutrition are not included in this book.

Donato Francesco Altomare
Maria Teresa Rotelli

Acknowledgments

Dr. Stefano Spinelli, Department of Emergency and Organ Transplantation, University Aldo Moro of Bari, Bari, Italy, has drawn most of the original figures.

Contents

Contributors

Donato Francesco Altomare, MD Department of Emergency and Organ Transplantation, University "Aldo Moro" of Bari, Bari, Italy

Patrizia Ancona, BSc Nutritionist Via Gramsci, Bari, Italy

Anna D'Eugenio, MD Consultant Nutritionist, Ambulatorio di Medicina Integrata, G. Bernabei Hospital, Ortona, Italy

Emanuele Felli, MD Department of General, Digestive and Endocrine Surgery, University Hospital of Strasbourg, Strasbourg, France

Antonio Finaldi, MD Centro Disturbi Alimentari Lucera Hospital (FG) ASL Foggia, Foggia, Italy

Pietro Genova, MD Department of Digestive, Hepatobiliary and Hepatic Transplantation Surgery, Henri Mondor University Hospital, Créteil, France

Dileep N. Lobo, MS, DM, FRCS, FACS, FRCPE Division of Gastrointestinal Surgery, Nottingham Digestive Diseases Centre, National Institute for Health Research Biomedical Research Unit, Queen's Medical Centre, Nottingham University Hospitals, Nottingham, UK

MRC/ARUK Centre for Musculoskeletal Ageing Research, University of Nottingham, Queen's Medical Centre, Nottingham, UK

Gennaro Martines, MD, PhD Surgical Unit, "M. Rubino"—Azienda Ospedaliero Universitaria Policlinico, Bari, Italy

Endocrinology Unit, Department of Emergency and Organ Transplantation, University Aldo Moro of Bari, Bari, Italy

Riccardo Memeo, MD, PhD Department of Emergency and Organ Transplantation, University Aldo Moro of Bari, Bari, Italy

John R. Nicholls, FRCS (Eng) Division of Surgery and Cancer, Imperial College London, London, UK

Sebastio Perrino, MD Endocrinology Unit, Department of Emergency and Organ Transplantation, University Aldo Moro of Bari, Bari, Italy

Arcangelo Picciariello, MD Department of Emergency and Organ Transplantation, University Aldo Moro of Bari, Bari, Italy

Filippo Pucciani, MD Department of Experimental and Clinical Medicine, University of Florence, Florence, Italy

Maria Teresa Rotelli, PhD Department of Emergency and Organ Transplantation, University Aldo Moro of Bari, Bari, Italy

Nilanjana Tewari, MBChB, MRCS, MSc Division of Gastrointestinal Surgery, Nottingham Digestive Diseases Centre National Institute for Health Research Biomedical Research Unit, Queen's Medical Centre, Nottingham University Hospitals, Nottingham, UK

Upper Gastrointestinal and General Surgery, University Hospitals Coventry and Warwickshire, Coventry, UK

Nilanjana Tewari and Dileep N. Lobo

1.1 Introduction

The metabolic and physiological changes associated with surgery are well established. There is a transient physiological state of impaired glucose tolerance (known as insulin resistance) along with the release of stress hormones and inflammatory mediators (cytokines, cortisol, catecholamines, glucagon) [1–3]. This results in attenuation of the anabolic effects of insulin, impairment of glucose uptake in peripheral tissues (e.g. skeletal muscle and adipose tissue) and enhancement of hepatic gluconeogenesis resulting in hyperglycaemia [4].

The regulation of food intake, energy storage and expenditure to maintain neutral balance involves complex homeostatic mechanisms. Information on the body's nutritional status and the presence of nutrients within the gastrointestinal (GI) tract are relayed via feedback mechanisms and then conveyed to the central nervous system (CNS). This initiates the transcription and release of neuropeptides in the hypothalamus and brainstem. Hormones derived from the GI tract and adipose tissue can act as appetite stimulants or suppressants (Fig. 1.1).

N. Tewari
Upper Gastrointestinal and General Surgery, University Hospitals Coventry and Warwickshire, Coventry, UK

Gastrointestinal Surgery, Nottingham Digestive Diseases Centre, National Institute for Health Research (NIHR) Nottingham Biomedical Research Centre, Nottingham University Hospitals NHS Trust and University of Nottingham, Queen's Medical Centre, Nottingham, UK

D. N. Lobo (✉)
Gastrointestinal Surgery, Nottingham Digestive Diseases Centre, National Institute for Health Research (NIHR) Nottingham Biomedical Research Centre, Nottingham University Hospitals NHS Trust and University of Nottingham, Queen's Medical Centre, Nottingham, UK

MRC/ARUK Centre for Musculoskeletal Ageing Research, University of Nottingham, Queen's Medical Centre, Nottingham, UK
e-mail: Dileep.Lobo@nottingham.ac.uk

© Springer Nature Switzerland AG 2019
D. F. Altomare, M. T. Rotelli (eds.), *Nutritional Support after Gastrointestinal Surgery*, https://doi.org/10.1007/978-3-030-16554-3_1

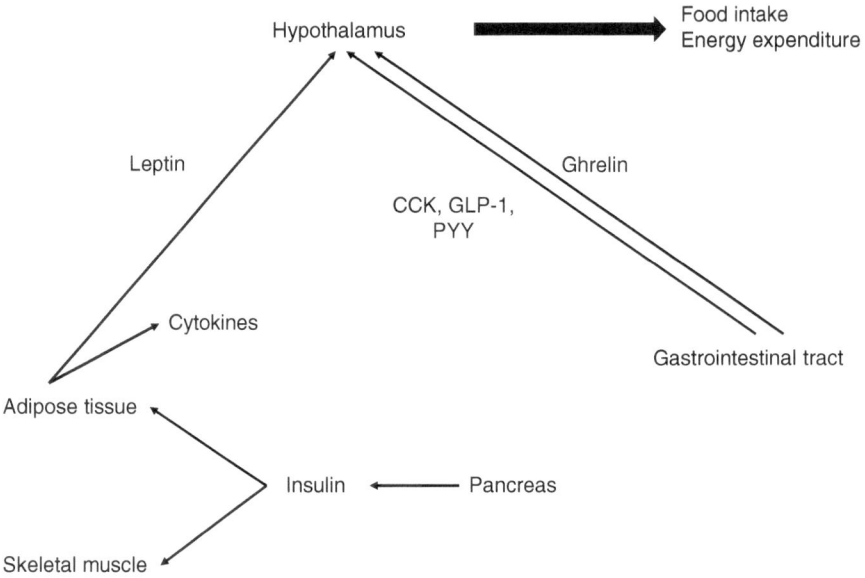

Fig. 1.1 Mechanisms involved in regulation of food intake and the gut-brain axis (*CCK* cholecystokinin, *GLP-1* glucagon-like peptide-1, *PYY* peptide YY)

Suboptimal nutritional status is believed to be associated with increased risk of postoperative complications. Patients whose food intake is reduced before hospital admission are more likely to suffer postoperative morbidity and mortality [5]. There are a number of factors which influence intake of food in the postoperative period, including metabolic and hormonal changes, reduced gastric and intestinal motility, ileus and pain.

Energy homeostasis is primarily controlled by the hypothalamus via several interconnecting nuclei (arcuate nucleus, paraventricular nucleus, ventromedial nucleus, dorsomedial nucleus and lateral hypothalamic area) (Fig. 1.2). Signals from the brain and humoral signals from the GI tract trigger the initiation and termination of meals. The appetite control centre is located in the arcuate nucleus of the hypothalamus [6, 7].

The autonomic nervous system with its parasympathetic and sympathetic divisions supplies innervation to the GI tract. Within the parasympathetic division are vagal and pelvic nerves, and the sympathetic division comprises splanchnic nerves. Auerbach's plexus contains a network of neurons which make up the enteric nervous system [8]. Sensors in the gut transmit signals to the CNS via afferent fibres in response to mechanical stimuli, nutrients in the gut lumen, neurohumoral stimuli (e.g. gut hormones), neurotransmitters and neuromodulators, as well as cytokines and inflammatory modulators produced by gut microbes [9].

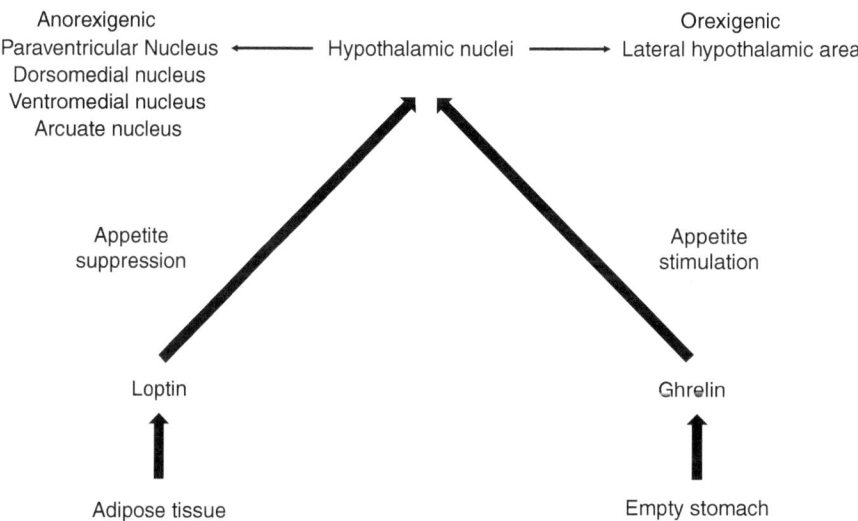

Fig. 1.2 Interaction between hypothalamic nuclei, gastrointestinal tract and adipose tissue in the control of food intake

Although mechanical distension of the stomach is an important mechanism involved in the regulation of appetite and food intake, the amount of food required for this exceeds that eaten in a typical meal [10]. Neural activity in the hippocampus and orbitofrontal cortex is stimulated by gastric distension. The signals are relayed via the vagus nerve and result in a sensation of satiety thereby leading to reduction in food intake [11, 12].

The crucial role played by the vagus nerve in responding to chemical stimuli (intragastric glucose), visceral mechanical stimuli (gastric distension) and hormonal signals such as leptin and ghrelin has been demonstrated in several studies. One such important study used functional MRI study to investigate forebrain activation induced by intragastric administration of glucose or monosodium glutamate in rats that had undergone total subdiaphragmatic vagotomy compared with control animals [13].

There is evidence from recent studies to suggest that energy intake is influenced both by the volume and type of food consumed. For example, protein appears to be more satiating than either fat or carbohydrate. Motility of the antrum-pylorus-duodenum complex also plays a role in the regulation of energy intake. A study in lean men showed that intraduodenal protein loading resulted in suppression of antral motility, stimulation of pyloric pressures and suppression of energy intake. The concentration of plasma cholecystokinin (CCK) and plasma glucagon-like peptide (GLP)-1 was directly related to the protein load administered [14]. Postoperatively, there are additional influences on food intake, such a gastric emptying and paralytic ileus, and these merit further research.

1.2   Leptin

Appetite, hunger and metabolism are affected by the 16 kDa hormone leptin. The obese gene coding leptin was first identified by positional cloning of ob/ob mice in 1994 [15]. The concentration of circulating leptin is proportional to the total amount of fat in the body, and it is synthesised mainly in white adipose tissue. Leptin receptors are highly expressed in the arcuate nucleus. When leptin binds to its receptors, Janus kinase 2—signal transducer and inhibition of adenosine monophosphate (AMP)-activated protein kinase (AMPK) activity is induced. When measured, circulating leptin levels show a diurnal and pulsatile pattern, and levels peak at night [16]. Leptin mediates its anorectic effects via the arcuate nucleus where it inhibits Neuropeptide Y/Agouti-related peptide (NPY/AgRP) neurons and activates proopiomelanocortin/cocaine- and amphetamine-regulated transcript (POMC/CART) neurons contributing to an increase in the expenditure of energy and lower food intake. Leptin is also produced in gastric epithelium and intensifies gut satiation signals, e.g. those that are mediated through CCK [17].

Leptin treatment was shown to help lower food intake in a study of diet-induced obese mice and, therefore, body weight [18]. A similar observational human study, in which nutrient intake and serum leptin concentrations were examined, showed an inverse correlation between leptin concentration and dietary intake in obese individuals (BMI > 30 kg/m^2); however there was no link between leptin and total dietary intake in those with BMI less than 25 kg/m^2 and those with BMI between 25 and 29.9 kg/m^2 [19]. Therefore, it was suggested that leptin may have an appetite-suppressing effect in obese individuals. However, the effects on caloric intake in lean and overweight people remain unclear. Obese individuals have been shown to be leptin-deficient because the hormone has been shown to affect control of appetite and body weight. However some data shows increased leptin concentrations in obese people, leading to what is known as 'leptin resistance'. That said, administering leptin to obese people has proven to be ineffective in aiding weight loss, and so the role of leptin in the control of food intake needs to be studied further.

1.3   Ghrelin

Ghrelin uses the hypothalamus to influence multiple brain regions involved in the food-appetite pathway, including the ventral tegmental area, nucleus accumbens, amygdala and hippocampus. The hormone regulates the activity and synaptic input organisation of midbrain dopamine neurons while promoting appetite. In rodents, it operates via the growth hormone secretagogue receptor (GHS-R) to increase food intake and has been shown to stimulate food intake in humans.

Ghrelin has also been shown to affect glucose homeostasis with a decrease of insulin secretion following intravenous administration of ghrelin, but insulin sensitivity was not changed in humans. In rodents, central and peripheral ghrelin administration causes dietary intake to increase as well as an increase in body weight along with a significant reduction in fat utilisation. The ghrelin receptor GHSR1a displays notable constitutive signalling activity, and so blockade by classical agonists cannot fully abolish ghrelin signalling. Therefore, central administration of a substance P derivate, an inverse agonist of the ghrelin receptor, was able to reduce food intake and body weight by decreasing the expression of NPY in the hypothalamus.

The P/D1 cells of the gastric fundus, the proximal small intestine and the epsilon cells of the pancreas generate ghrelin. It is the only known orexigenic gut hormone and in humans; it has been shown to increase food intake. It helps decrease insulin secretion and increase gastrointestinal motility. It is shown to regulate short-term feeding control and long-term body weight. In a study of rats, ghrelin administration was found to enhance gastric motility and induce excitation of neurons in the lateral septum [20]. Agonists of the ghrelin pathway have been developed, such as growth hormone secretagogue receptor antagonists (GHR-P6, JMV 2959, YIL-781 and YIL-870), ghrelin analogue BIM-28163 which binds to but does not activate the GHS-1a receptor and ghrelin receptor agonist GSK1614343. In mice, some ghrelin agonists have been shown to improve appetite [21]. In one study of 25 patients undergoing oesophagectomy for cancer, circulating ghrelin concentrations on day 7 postoperatively were found to be 38.7% lower than those measured preoperatively. At 36–80 months postoperatively, plasma ghrelin concentrations were restored to preoperative values. A correlation between postoperative reductions in BMI and reduced plasma ghrelin was observed early (6–24 months), but this was not seen at 36 months postoperatively [22]. In another study of 16 overweight and obese adults (BMI 29.8 ± 2.9 kg/m^2), who were fed isocaloric meals that differed in fat and carbohydrate content, ghrelin concentrations were linked with hunger, as reported by subjective appetite measures and food intake [23]. Another study investigated the use of ghrelin infusions in the postoperative period following open colorectal surgery and demonstrated increased gastric emptying [24], suggesting that ghrelin agonists may have the potential to be therapeutic in early re-establishment of oral feeding and the management of postoperative ileus.

1.4 Insulin

The brain has insulin receptors across it, particularly in the hypothalamic nuclei, which are known to affect the regulation of dietary intake. Links have been demonstrated between increased plasma insulin concentrations and changes in brain activity in the anterior cingulate cortex, in the orbitofrontal cortex, in the sensorimotor cortex and in the hypothalamus. In rodent and baboon experiments, insulin has been shown to have anorectic effects, if administered centrally [25].

Like leptin, insulin is thought to have a lipostatic role, and increased adiposity is associated with a state of insulin resistance [26]. In a study on the effects of insulin administration on food intake, two groups of rats were fed with a high-fat diet (HFD) to create diet-induced obesity (DIO) or a low-fat chow diet (LFD). Half the HFD rats were crossed over to LFD after 22 weeks. Insulin was then administered centrally (directly into the third ventricle), and this was related to reduced food intake and weight loss in the LFD group and the group which changed from HFD to LFD but not in the fully HFD rats. Thus, insulin administration was associated with reduced food intake, but the effects were not seen in DIO rats [27]. Carbohydrate treatment prior to surgery preserved intestinal barrier function and led to increased food intake in an intestinal ischaemia-reperfusion rat model. Improved food intake was believed to be due to preserved intestinal integrity and/or attenuation of insulin resistance [28]. Studies related to the control of eating behaviours associated with eating disorders have all flagged leptin, insulin and ghrelin as of some use. These hormones are all linked to long-term energy status, but there are some other gut hormones, such as cholecystokinin (CCK), glucagon-like peptide 1 (GLP-1), pancreatic polypeptide (PP), amylin and peptide YY (PYY) which are also thought to have a crucial role in triggering and stopping food intake.

1.5   Cytokines

The release of cytokines such as tumour necrosis factor-α (TNFα) and interleukin 6 (IL-6) causes an inflammatory state, which characterises the postoperative period [29]. Links have been shown between several cytokines and the regulation of food intake, and these may be important postoperatively. A cross-sectional study of Japanese men showed circulating IL-1b concentrations helped improve the rate of eating [30]. In another study in Japanese men, plasma IL-1b and IL-6 concentrations were shown to be positively associated as determined indirectly by the homeostatic model assessment [31]. Gene knockout mice were studied to examine the tumour growth factor family cytokine MIC-1/GDF15. MIC-1/GDF15 gene knockout mice weighed more and had increased adiposity, signalling greater spontaneous food intake. Administering human recombinant MIC-1/GDF15 caused serum levels to increase in these mice to within the normal human range and also decreased food intake and body weight. These results point to a relationship between MIC-1/GDF15 and the physiological regulation of appetite and energy storage [32].

1.6   Gut Microbiota

Gut microbiota and bile acids have been shown to regulate energy metabolism in multiple studies. They have been shown to positively affect bariatric surgery. It is thought that they act in metabolism through a complicated network of communication relating to the vagus nerve and neuroendocrine pathways [33]. There is evidence that obese individuals show differences in the composition of gut microbiota, which can be altered in response to changes in caloric consumption and the makeup of macronutrients [34]. Recent work demonstrates that bariatric surgery is associated with increased microbial diversity and altered microbial composition following Roux-en-Y gastric bypass (RYGB) in humans and following sleeve gastrectomy (SG) and RYGB in rodents [35]. It was also shown that colonisation of germ-free mice with faecal material from mice after gastric bypass leads to decreased body weight and adiposity [36]. This supports evidence that RYGB-associated gut micro-biota can have direct effects on host metabolism. A recent human study also measured the faecal microbial makeup and metabolites of normal weight and morbidly obese pre-bariatric surgery patients, as well as individuals following either RYGB or LAGB. It demonstrated that there was a significant change in the faecal microbiome and metabolome in patients who underwent RYGB and AGB compared to normal weight and obese participants. Since the two bariatric procedures are anatomically different, it is expected that the outcomes were more pronounced in patients undergoing gastric bypass compared to LAGB [37].

1.7 Metabolic Surgery

As mentioned earlier, the way appetite is regulated entails interactions between nutrients, gut-derived hormones [such as peptide YY (PYY) and GLP-1] and the brain. These hormones, released secondary to gastric distension, digestion of food and via neuronal signals, regulate feelings of satiety by decreasing hypothalamic orexigenic signalling and increasing anorectic signalling. They also mediate feedback mechanisms on the gastrointestinal tract by contributing to prolonged gastric distension thereby increasing satiety between meals. In health, these mechanisms act to facilitate the control of food intake and gastrointestinal transit [38]. As a result of multiple studies, we know that decreased basal and postprandial concentrations of PYY and GLP-1 in obese compared with lean individuals result in decreased feelings of satiety, higher levels of food intake and obesity. Metabolic (bariatric) surgery is still the most effective treatment option for obesity, with sustained long-term weight loss, remission of obesity-related comorbidities and improved life expectancy as outcomes [39]. Treatments such as the RYGB procedure (where the stomach and proximal intestine are bypassed with enhanced nutrient delivery to the distal intestine) and SG (removal of 80–90% of the stomach leading to rapid emptying of gastric contents into the small intestine) result in long-term excess weight loss of 50–70%. However, the metabolic and anorectic effects of RYGB are evident within a few days of surgery, prior to the occurrence of significant weight loss, and provide a valuable mechanistic insight into the gut regulation of food intake. Increased GLP-1 and PYY is one of the mechanisms by which RYGB induces weight loss [40]. It has recently been demonstrated that within 2 weeks of RYGB, and independent of calorie restriction, there is increased postprandial release of GLP-1 and PYY. A recent randomised study demonstrated that postprandial GLP-1, PYY and insulin responses may be further enhanced, and ghrelin concentrations decreased markedly, if concomitant gastric fundus resection was performed with RYGB, compared with standard RYGB in which the gastric fundus is left in situ [41]. A human study carried out recently showed changes in gut hormones after a sleeve gastrectomy. Most interestingly, substantially decreased fasting and postprandial concentrations of ghrelin, amylin and leptin were seen, with significant postprandial increases in PYY and GLP-1 concentrations. This implies that the weight loss following sleeve gastrectomy was not only from the restrictive element but also from positive hormonal effects [42]. Results of rodent and human studies point to further roles for central PYY expression and central melanocortin signalling (via the melanocortin-4 receptor) in the post-RYGB regulation of energy balance and, so, weight loss [43, 44]. Another recent functional MRI study showed alterations in neural responses to food cues after RYGB, with postsurgical reductions in brain activation in response to food cues that were high (versus low) in caloric density [45]. However, these alterations in neuronal responses were not present after sleeve gastrectomy. This was shown in a rodent study in which animals treated by sleeve gastrectomy kept their preoperative food preferences [46]. Following RYGB, the pathophysiological state of leptin resistance, which characterises obese individuals, appeared to be normalised [47]. There was also substantial correlation between leptin and adiponectin gene expression in visceral fat. This study suggested another potential mechanism for regulation of food intake because leptin is reported to inhibit AMPK activity in the hypothalamus. That said, administering leptin following RYGB did not promote further weight loss in a randomised double-blind study [48].

1.8   Conclusion

Given the worldwide epidemic of obesity, it is clinically important to fully comprehend factors involving the regulation of appetite and nutrient intake as well as both the long-term and temporary anorectic effects of metabolic and major surgery, respectively. Despite several neurohormonal peptides being identified as contributing to food intake, many such as CCK, PP and amylin have shown to have an impact on animals, but this has not been replicated in human studies. That said, with an increase in the popularity of metabolic surgery, it is expected that the mechanisms controlling appetite in humans will be further examined. The mechanisms underlying long-term regulation of appetite after metabolic surgery and the development of neuromodulator analogues need to be further researched, the outcome of which will be valuable in efforts to control obesity.

References

1. Nygren J, Thorell A, Efendic S, et al. Site of insulin resistance after surgery: the contribution of hypocaloric nutrition and bed rest. Clin Sci (Lond). 1997;93:137–46.
2. Thorell A, Loftenius A, Andersson B, et al. Postoperative insulin resistance and circulating concentrations of stress hormones and cytokines. Clin Nutr. 1996;15:75–9.
3. Akintola DF, Sampson B, Burrin J, et al. Changes in plasma metallothionein-1, interleukin-6, and C-reactive protein in patients after elective surgery. Clin Chem. 1997;43:845–7.
4. Bisgaard T, Kristiansen VB, Hjortso NC, et al. Randomized clinical trial comparing an oral carbohydrate beverage with placebo before laparoscopic cholecystectomy. Br J Surg. 2004;91:151–8.
5. Kuppinger D, Hartl WH, Bertok M, et al. Nutritional screening for risk prediction in patients scheduled for abdominal operations. Br J Surg. 2012;99:728–37.
6. Perry B, Wang Y. Appetite regulation and weight control: the role of gut hormones. Nutr Diabetes. 2012;2:e26.

7. Yu JH, Kim MS. Molecular mechanisms of appetite regulation. Diabetes Metab J. 2012;36:391–8.
8. Langley J. The autonomic nervous system. Cambridge: Part I; 1921.
9. Janig W, Morrison JF. Functional properties of spinal visceral afferents supplying abdominal and pelvic organs, with special emphasis on visceral nociception. Prog Brain Res. 1986;67:87–114.
10. Ritter RC. Gastrointestinal mechanisms of satiation for food. Physiol Behav. 2004;81:249–73.
11. Wang GJ, Yang J, Volkow ND, et al. Gastric stimulation in obese subjects activates the hippocampus and other regions involved in brain reward circuitry. Proc Natl Acad Sci U S A. 2006;103:15641–5.
12. Niijima A. Reflex effects of oral, gastrointestinal and hepatoportal glutamate sensors on vagal nerve activity. J Nutr. 2000;130:971S–3S.
13. Tsurugizawa T, Uematsu A, Nakamura E, et al. Mechanisms of neural response to gastrointestinal nutritive stimuli: the gut-brain axis. Gastroenterology. 2009;137:262–73.
14. Ryan AT, Feinle-Bisset C, Kallas A, et al. Intraduodenal protein modulates antropyloroduodenal motility, hormone release, glycemia, appetite, and energy intake in lean men. Clin Nutr. 2012;96(3):474–82.
15. Zhang Y, Proenca R, Maffei M, et al. Positional cloning of the mouse obese gene and its human homologue. Nature. 1994;372:425–32.
16. Minokoshi Y, Alquier T, Furukawa N, et al. AMP-kinase regulates food intake by responding to hormonal and nutrient signals in the hypothalamus. Nature. 2004;428:569–74.
17. Munzberg H. Leptin-signaling pathways and leptin resistance. Forum Nutr. 2012;63: 123–32.
18. Knight ZA, Hannan KS, Greenberg ML, et al. Hyperleptinamia is required for the development of leptin resistance. PLoS One. 2010;5:e11376.
19. Nakamura Y, Ueshima H, Okuda N, et al. Serum leptin and total dietary energy intake: the INTERLIPID study. Eur J Nutr. 2012;52(6):1641–8.
20. Gong Y, Xu L, Guo F, et al. Effects of ghrelin on gastric distension sensitive neurons and gastric motility in the lateral septum and arcuate nucleus regulation. J Gastroenterol. 2014;49(2):219–30.
21. Hassouna R, Labarthe A, Zizzari P, et al. Actions of agonists and antagonists of the ghrelin/GHS-R pathway on GH secretion, appetite, and cFos activity. Front Endocrinol (Lausanne). 2013;4:25.
22. Miyazaki T, Tanaka N, Hirai H, et al. Ghrelin level and body weight loss after esophagectomy for esophageal cancer. J Surg Res. 2012;176:74–8.
23. Gibbons C, Caudwell P, Finlayson G, et al. Comparison of postprandial profiles of ghrelin, active GLP-1, and total PYY to meals varying in fat and carbohydrate and their association with hunger and the phases of satiety. J Clin Endocrinol Metab. 2013;98(5):E847–55.
24. Falken Y, Webb DL. Abraham-Nordling et al. Intravenous ghrelin accelerates postoperative gastric emptying and time to first bowel movement in humans. Neurogastroenterol Motil. 2013;25(6):474–e364.
25. Air E, Benoit SC, Blake Smith KA, et al. Acute third ventricular administration of insulin decreases food intake in two paradigms. Pharmacol Biochem Behav. 2002;72(1-2):423–9.
26. Horakova D, Stejskal D, Pastucha D, et al. Potential markers of insulin resistance in healthy vs obese and overweight subjects. Biomed Pap Med Fac Univ Palazky Olomouc Czech Repub. 2010;154:245–9.
27. Begg DP, Mul JD, Liu M, et al. Reversal of diet-induced obesity increases insulin transport into cerebrospinal fluid and restores sensitivity to the anorexic action of central insulin in male rats. Endocrinology. 2013;154:1047–54.
28. Luttikhold J, Oosting A, van den Braak CC, et al. Preservation of the gut by preoperative carbohydrate loading improves postoperative food intake. Clin Nutr. 2012;32(4): 556–61.
29. Kragsbjerg P, Holmberg H, Lee IC, et al. Adult-onset PYY overexpression in mice reduces food intake and increases lipogenic capacity. Neuropeptides. 2012;46:173–82.

30. Mochizuki K, Misaki Y, Miyauchi R, et al. A higher rate of eating is associated with higher circulating interleukin-1beta concentrations in Japanese men not being treated for metabolic diseases. Nutrition. 2012;28:978–83.
31. Misaki Y, Miyauchi R, Mochikuzi K, et al. Plasma interleukin-1beta concentrations are closely associated with fasting blood glucose levels in healthy and preclinical middle-aged nonoverweight and overweight Japanese men. Metabolism. 2010;59:1465–71.
32. Tsai VW, Macia L, Johnen H, et al. TGF-b Superfamily Cytokine MIC-1/GDF15 is a physiological appetite and body weight regulator. PLoS One. 2013;8:e55174.
33. Forsythe P, Bienenstock J, Kunze WA. Vagal pathways for microbiome-brain-gut axis communication. Adv Exp Med Biol. 2014;817:115–33.
34. Parks BW, Nam E, Org E, et al. Genetic control of obesity and gut microbiota composition in response to high-fat, high-sucrose diet in mice. Cell Metab. 2013;17:141–52.
35. Zhang H, DiBaise JK, Zuccolo A, et al. Human gut microbiota in obesity and after gastric bypass. Proc Natl Acad Sci U S A. 2009;106:2365–70.
36. Liou AP, Paziuk M, Luevano JM Jr, et al. Conserved shifts in the gut microbiota due to gastric bypass reduce host weight and adiposity. Sci Transl Med. 2013;5:178ra41.
37. Ilhan ZE, DiBaise JK, Isern NG, et al. Distinctive microbiomes and metabolites linked with weight loss after gastric bypass, but not gastric banding. ISME J. 2017;11:2047–58.
38. Bewick GA. Bowels control brain: gut hormones and obesity. Biochem Med (Zagreb). 2012;22:283–97.
39. Sjostrom L, Peltonen M, Jacobson P, et al. Bariatric surgery and long-term cardiovascular events. JAMA. 2012;307:56–65.
40. Pournaras DJ, Osborne A, Hawkins SC, et al. The gut hormone response following Roux-en-Y gastric bypass: cross-sectional and prospective study. Obes Surg. 2010;20:56–60.
41. Chronaiou A, Tsoli M, Kehagias I, et al. Lower ghrelin levels and exaggerated postprandial peptide-YY, glucagon-like peptide-1 and insulin responses after gastric fundus resection in patients undergoing Roux-en-Y gastric bypass: a randomised clinical trial. Obes Surg. 2013;22:1761–70.
42. Dimitradis E, Daskalakis M, Kampa M, et al. Alterations in gut hormones after laparoscopic sleeve gastrectomy: a prospective clinical and laboratory investigational study. Ann Surg. 2013;257:647–54.
43. Gelegen C, Chandarana K, Choudhury AI, et al. Regulation of hindbrain Pyy expression by acute food deprivation, prolonged calorie restriction and weight loss surgery. Am Physiol Endocrinol Metab. 2012;303:E659–68.
44. Hatoum IJ, Stylopoulos N, Vanhoose AM, et al. Melanocortin-4 receptor signaling is required for weight loss after gastric bypass surgery. J Clin Endocrinol Metab. 2012;97:E1023–31.
45. Ochner CN, Kwok Y, Conceicao E, et al. Selective reduction in neural responses to high calorie foods following gastric bypass surgery. Ann Surg. 2011;253:502–7.
46. Saeidi N, Nestoridi E, Kucharczyk J, et al. Sleeve gastrectomy and Roux-en-Y gastric bypass exhibit differential effects on food preferences, nutrient absorption and energy expenditure in obese rats. Int J Obes (Lond). 2012;36:1396–402.
47. Chen J, Pamuklar Z, Spagnoli A, et al. Serum leptin levels are inversely correlated with omental gene expression of adiponectin and markedly decreased after gastric bypass surgery. Surg Endosc. 2012;26:1476–80.
48. Korner J, Conroy R, Febres G, et al. Randomized double-blind placebo-controlled study of leptin administration after gastric bypass. Obesity. 2013;21(5):951–6.

Nutritional Support After Surgery of the Esophagus

Pietro Genova and Antonio Finaldi

2.1 Introduction

Surgery of the esophagus may be indicated for benign and malignant pathologies. Esophageal cancer represents the most frequent indication for surgical treatment, and the important nutritional concern arisen in relationship between cancer, surgery, and nutritional status of operated oncological patients has been explored by a vast scientific literature. Among benign pathologies indicating surgery involving the esophagus, achalasia has been largely studied, with surgery representing one of several effective therapeutic options.

2.2 Esophageal Cancer

Esophageal resections are invasive and demanding surgical procedures. The most frequent indication is esophageal cancer (83%). Other indications include caustic injury (8%), functional disorders (2%), traumatic injury (2%), benign tumors (1%), and other causes, such as postoperative complications (4%) [1].

By simplifying the current debate on indications, esophagectomy is considered the standard approach for resectable esophageal cancers and esophagogastric junction malignancies. It represents the first-line treatment for cancers staged T1b or T2N0M0, and it is considered as a second-line treatment after chemotherapy or chemoradiotherapy for locally advanced cancer (tumors staged T3-T4a or N+ and M0) [2].

The most frequent surgical techniques are (Fig. 2.1) Ivor-Lewis transthoracic esophagectomy (i.e., abdominal and thoracic access with intrathoracic anastomosis,

P. Genova (✉)
Department of Digestive, Hepatobiliary and Hepatic Transplantation Surgery, Henri Mondor University Hospital, Créteil, France

A. Finaldi
Centro Disturbi Alimentari Lucera Hospital (FG) ASL Foggia, Foggia, Italy

© Springer Nature Switzerland AG 2019
D. F. Altomare, M. T. Rotelli (eds.), *Nutritional Support after Gastrointestinal Surgery*, https://doi.org/10.1007/978-3-030-16554-3_2

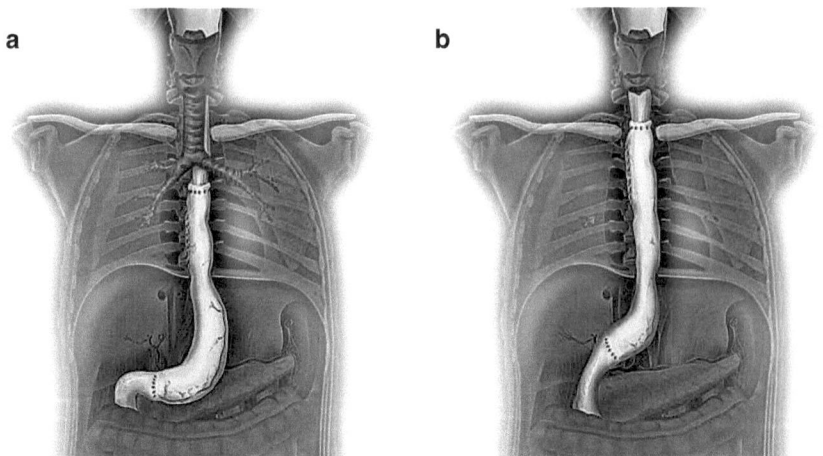

Fig. 2.1 Types of esophagectomy: (**a**) Ivor-Lewis transthoracic esophagectomy (abdominal and thoracic access with intrathoracic anastomosis, over 70%); (**b**) McKeown tri-incisional esophagectomy (abdominal, thoracic, and cervical access with cervical anastomosis, 12%). Transhiatal esophagectomy (abdominal and cervical access with cervical anastomosis, 4%, not represented) requires a cervical anastomosis like McKeown esophagectomy, from which it differs for the double access instead of the triple one. A gastric conduit is represented

performed in over 70% of cases), McKeown tri-incisional esophagectomy (i.e., abdominal, thoracic, and cervical access with cervical anastomosis, 12%), and tran-shiatal esophagectomy (i.e., abdominal and cervical access with cervical anastomosis, 4%) [1]. A minimally invasive approach is performed in approximately 10% of patients. The conduit used to restore digestive tube continuity is represented by the stomach in 90% of patients, the colon in 6%, and the jejunum in 4% [3].

Despite recent improvements in preoperative prehabilitation, surgical techniques, and perioperative care, esophageal resections are still associated with high mortality and morbidity rates. For oncological resections, the median postoperative mortality ranges from 4.9% to 18% [4]. The 5-year survival rate varies from 17% to 22% [5], depending on whether surgery is performed in low- or high-volume centers. Overall postoperative complications occur in 50–60% of patients [6], with the most frequent complication being anastomotic leakage (up to 35% of patients), pulmonary complications (up to 38.9%), recurrent laryngeal injury (up to 31.1%), chyle leak (up to 11%), and functional disorders (up to 50%) [7, 8].

Many studies have confirmed that esophageal resections have a significant effect on operated patients' nutritional status, with a significant impact on their postoperative outcomes and quality of life [3, 9–11].

Therefore, it is of major importance to implement the best available strategies in order to meet optimal nutritional requirements before and after esophageal surgery. This chapter will provide an overview of the postoperative management of the patient's nutritional status after esophagectomy.

2.3 Achalasia

Regarding esophageal benign pathology representing an indication for surgery, a particular attention should be paid to achalasia. Achalasia is a motility disorder caused by insufficient lower esophageal sphincter (LES) relaxation and loss of esophageal peristalsis. Its clinical presentation is characterized by dysphagia to solids and liquids and regurgitation of saliva and bland undigested food, frequently accompanied by substernal chest pain, weight loss, and pyrosis [12].

Therapeutic options for achalasia aim at reducing LES hypertonia and include pharmacologic, endoscopic, and surgical approaches. Regarding surgery, laparoscopic Heller cardiomyotomy with gastric fundoplication represents the preferred treatment modality in many centers [13]. It is indicated for good candidates for surgery, as an alternative treatment to endoscopic treatments or after these latter ones have failed [12, 14]. The Heller cardiomyotomy consists of an anterior muscle fiber longitudinal incision extended 6 cm into the esophagus and 2–3 cm into the stomach as measured from the gastroesophageal junction, without disrupting the mucosal layer [14, 15]. A gastric fundoplication (according to Dor, Toupet, or Nissen techniques) is generally performed to prevent gastroesophageal reflux (Fig. 2.2) [12, 16, 17]. Symptom relief is reported in 90% of operated patients at 24 postoperative months [13] although efficacy appears to decrease over time, reaching 57% 6 years after surgery [18]. A postoperative gastroesophageal reflux disease (GERD) is reported in 17.5% of patients [13]. A more invasive esophagectomy may be proposed to good surgical candidates in case of persistent or recurrent achalasia and failure of previous less invasive treatments, especially in the presence of a megaesophagus and of significant esophageal dilation and tortuosity [12, 14].

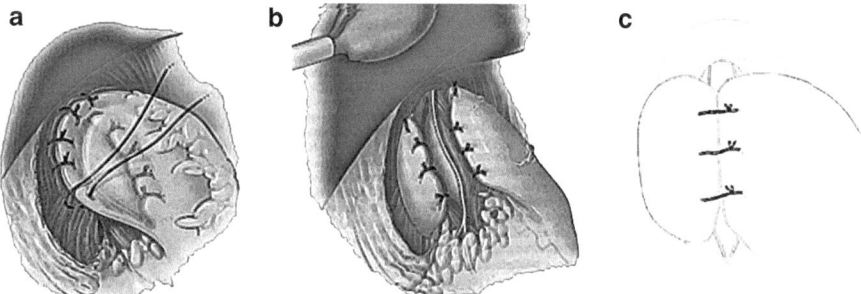

Fig. 2.2 Heller cardiomyotomy with anti-reflux fundoplication: (**a**) Dor technique, (**b**) Toupet technique, and (**c**) Nissen technique

2.4   Nutritional Status After Esophagectomy for Cancer

Nutritional status after esophageal surgery has been largely explored for patients affected by cancer. Notably, a systematic review published in 2016 and concerning the nutritional impact of esophagectomy has reported that about 98% of patients included were oncological [3].

Surgery represents a iatrogenic injury activating a metabolic stress response, characterized by the release of stress hormones and inflammatory mediators stimulating the catabolism of glycogen, fatty acids, and proteins. Metabolic substrates are diverted from their normal peripheral purposes, leading to the reduction of muscle mass, impairment of wound healing, and inhibition of immune response. That impacts negatively on short- and long-term surgical outcomes and makes appropriate nutritional support a necessary practice [19].

Malnutrition, defined according to the European Society of Enteral and Parenteral Nutrition (ESPEN) guidelines [19, 20], is a frequent condition in patients with esophageal cancer, occurring in up to 78.9% of patients [21]. In the preoperative period, several factors may affect the patient nutritional status: the catabolism associated to cancer-related chronic inflammation, the dysphagia, the loss of appetite and the weight loss, or the effects of neoadjuvant treatments [10, 11].

The postoperative nutritional status is influenced by the extent of the esophageal resection, the severity of the consequent inflammatory response, and the occurrence of complications and functional disorders, which may persist in the long-term and significantly impair the patient's quality of life [11]. During hospital stay, several complications may have an impact on the nutritional status by making feeding routes impractical. Anastomotic leakage, aspiration risk related to swallowing dysfunction, and gastrointestinal motility disorders often require a prolonged *nihil* per os regimen. Intolerance to enteral nutrition may be registered also when it is provided by enteral feeding tubes, whereas parenteral nutrition may become the only available feeding route in case of severe complications like chylothorax [11]

Functional disorders such as anastomotic stricture, accelerated or delayed gastric emptying, dumping syndrome, and gastroesophageal reflux occur in about 50% of patients undergoing esophagectomy with gastric tube reconstruction [8]. The most frequent symptoms reported 1 year after esophageal resections are inappetence, early satiety, dumping syndrome, dysphagia, and regurgitation [22]. During the same period, most patients report an altered stool frequency, with negative impact on social life [22]. Gastrointestinal symptoms may be persistent, being reported also after 10 years from surgery in long-term survivors [23]; among these, swallowing dysfunction appears to have the strongest correlation with a compromised global health status [24].

Risk factors associated with postoperative weight loss include high preoperative BMI, female sex, neoadjuvant treatment, and gastrointestinal symptoms [3, 25, 26]. In addition to postoperative functional disorders, malabsorption may play an important role in weight loss and nutritional status impairment observed after esophagectomy. Its etiology would be multifactorial, involving exocrine pancreatic insufficiency (EPI), bile acid malabsorption (BAM), and small intestinal bacterial overgrowth (SIBO).

EPI would be associated to vagal denervation and loss of endogenous neuroendocrine signals. BAM would result from vagal denervation and disruption of enterohepatic circulation of bile acids. SIBO would follow the altered intestinal motility, diminished gastric acid secretion, and impaired intestinal secretions [27–30].

At 12 months after surgery, energy and protein intake is found to be 24% and 7% below the recommended values, respectively [31]. Only 7% of the operated patients have a sufficient intake of all micronutrients. Suboptimal intake of micronutrients is often reported for folic acid (85%), vitamin D (61%), calcium (49%), and vitamin B1 (48%) [31]. Weight loss after surgery may be severe (>15% of preoperative weight in 33.8% and 36% of patients after 3 and 5 years, respectively) and long-lasting (≥10 years) [23, 25]. In a Swedish nationwide prospective study published in 2007, at 6 months after esophagectomy, 63.7% of patients had lost >10% of their preoperative BMI, and 20.4% of them had lost >20%. A systematic review on nutritional effects of esophagectomy published in 2016 has confirmed that more than half of patients undergoing esophageal resections loose >10% of their preoperative weight between 6 and 12 months after surgery [3]. A plateau in weight loss is registered afterward, but 27–95% of operated patients fail to recover their preoperative condition after surgery [3].

However, despite the abundance of studies on the nutritional status and esophageal surgery, the literature lacks of research specifically focused on nutritional inadequacy and postoperative dietary interventions [11].

At discharge, it has been estimated that oral intake of energy and nitrogen are at 70% and 65% of the nutritional requirements, respectively [32]. Limited oral intake and necessity of nutritional support after discharge are reported in 8–48% of the operated patients [3, 22, 32]. It has been described a mainly overnight feeding, providing 1000 kcal daily, with 2% and 1% of patients still requiring artificial EN at 6 and 12 months after surgery, respectively [22].

More recently, because of the frequent and persistent nutritional weight loss reported after esophagectomy, it has become usual to discharge patients with enteral feeding tubes, in order to provide a prolonged nutritional support. The preferred feeding route is jejunostomy, whose placement is common (68% in England) [33] and which appears to be more tolerated than nasoduodenal or nasojejunal feeding tubes. In this regard, a retrospective study on routine use of jejunostomy after esophagectomy published in 2017 has reported some useful data. Notably, it concludes that patients are able to maintain their weight during tube feeding but not after nutritional support withhold, independently of the duration of artificial nutrition. The median uninterrupted period of artificial EN following surgery was 35 days (range 13 74). Tube feeding continued over 180 days in 4% of patients and until death in 6%. In the multivariate analysis, independent preoperative factors associated to tube feeding duration of 35 days or more were the American Society of Anesthesiologists (ASA) score >2, the presence of systemic comorbidities, neoadjuvant chemoradiation therapy, and discharge with home tube feeding regimen. Artificial tube feeding was restarted in 15% patients within 6 months from surgery, and the most frequently reported indication was insufficient oral intake (56% of cases). Artificial EN was restarted via jejunostomy in 61% of patients at 34 days

[26] and via nasojejunal tube because of previous jejunostomy withdrawal in the remaining 39% of cases at 98 days [16].

It has been suggested that the use of immune-nutrition in surgical patients may have a positive modulatory effect on the inflammation and immune responses, by ameliorating gut functions, notably after gastrointestinal cancer resections [34–38].

In conclusion, the management of the nutritional status of patients undergoing esophageal resection appears to be a broad and complex issue, which requires an active collaboration between all actors involved: patients, surgeons, dieticians, and caregivers.

2.5   Achalasia, Surgery, and Nutritional Considerations

Differently from esophagectomy for cancer, a very limited literature has focused on achalasia, its surgical treatment, and nutrition.

2.5.1   Nutritional Status Associated to Achalasia

Current literature reports a very limited number of studies exploring nutritional implications of achalasia and nutritional effects of surgical procedures performed to treat this pathology [39, 40].

A retrospective cross-sectional study of patients with achalasia has shown that they were overweight or obese in 70% of cases, but 26% were at moderate or high nutrition risk according to the Malnutrition Universal Screening Tool (MUST) [41]. In a prospective study including only 19 patients with achalasia, dietary alterations due to dysphagia have been reported in 80% of cases and consuming food less than usual in 90% [42]. Furthermore, 80% of patients have shown relevant weight loss over the course of 6 months before diagnosis, and in 75% of cases low prealbumin levels have been reported, demonstrating poor nutrition [43]. In a retrospective study including 100 patients with achalasia undergoing pneumatic dilatation or Heller myotomy, weight loss has been reported in 51% of cases, with a median weight loss of 12.7 kg (range 6.3–18.1) [39]. According to high-resolution manometry (HRM) findings, the Chicago Classification of Esophageal Motility Disorders distinguishes three types of achalasia [43]. Patients reporting weight loss have shown a type II achalasia in 71% of cases. Particularly, weight loss has been reported in 63% of patients with type II achalasia, resulting absent in 73% of patients with type III. Only 32 patients have been included for postoperative nutritional evaluation. During a median follow-up period of 22 months (range 6–90 months), 43% of patients initially reporting weight loss have denied regaining their weight after treatment. However, there were no differences in the modified achalasia dysphagia score (MADS) between patients with or without weight gain after intervention, which suggests no difference in dysphagia symptoms between the two groups [39].

It remains unclear the reason why some patients with achalasia report a significantly greater weight loss on clinical presentation compared to others, despite the presence of a mechanical obstruction in both groups. Notably, patients with weight loss report symptoms for a significantly shorter period (12 vs. 24 months), which suggest a lack of dietary adaptation in the early pathological process. Moreover, it is unclear also why type II achalasia patients appear to report weight loss more frequently, considering that dietary modifications related to swallowing difficulty would represent the main mechanism for weight loss in all types of achalasia. Furthermore, for the same reasons, a postoperative weight regains after invasive intervention on LES would be expected, but a high rate of patients does not confirm this hypothesis. That would suggest that it is unlikely that LES physiology is the sole etiology of weight loss in achalasia. In this context, a potential explanation could be that patients with type II achalasia might have a different pro-inflammatory response compared to other subtypes of achalasia, with more significant catabolism and alterations in gastrointestinal hormones mediating energy homeostasis or suppress appetite [39, 44].

2.5.2   Nutrition After Surgery for Achalasia

In the literature, nutritional support after surgery for achalasia is not the objective of specific studies, and data are very limited. However, it remains useful to give an overview of the general attitude described. A leak test is most frequently performed to rule out a mucosal perforation, and it is commonly carried out on the first postoperative day through a water-soluble contrast swallow [15, 45, 46]. Some surgeons perform the leak test intraoperatively, by using a methylene blue-stained solutions administered via a nasogastric tube [47]. If no perforation is shown, a liquid diet is started the same day. The patient is generally discharged from hospital after an average of 3 days following surgery [13]. Patients are generally suggested to progress gradually in oral intake recovery, advancing as tolerated [48]. However, some institutions suggest a liquid diet for the first week after surgery [49], and generally a soft food diet is recommended for 2–4 weeks [50, 51]. Moreover, since operated patients may experience heartburn, hygienic-dietary recommendations to reduce gastro-esophageal reflux can be useful.

2.6   Nutritional Postoperative Support After Surgery of the Esophagus

After partial or total removal of the esophagus, the transit time in the upper digestive tube is severely impaired because of altered peristaltic activity and abnormal control of food passage. Among the most common symptoms, reflux and dumping syndrome are to be noted.

An important rule to reduce dismotility-related symptoms is represented by eating slowly and chewing food completely. Therefore, the bolus should be swallowed when it is almost liquid, thanks to the chewing process and the salivation, and in small quantities.

Bites should be small and the overall portion of food moderate. The overall daily intake of food should be divided into six meals, and the last one should be taken at least 3 hours before going to bed. In any case, patients should avoid lying down soon after the meal.

It is generally suggested to limit drinking during meals. Therefore, liquids (water, juices, energy drinks) should be taken between two consecutive meals. Patients may take a little quantity of liquids or add some sauce, cream, or yogurt in order to make food softer when salivation alone is insufficient.

The patient should always pay attention to salivation in order to start food digestion already in the mouth, thanks to the action of salivary enzymes (such as amylase), and of the mucus, facilitating food transit thanks to its content of glucosaminoglicans and glycoproteins.

During the first postoperative weeks, it is advisable to eat foods that can be crushed, passed, or reduced to pulp, the so-called soft diet.

It is better to avoid mixing foods since this method incorporates air into the nutrient solutions, which may cause abdominal bloating, quicker sense of fullness, and higher risk of regurgitation.

The caloric intake must be higher than normal because of the weight loss due to underlying pathology, surgical stress, and impaired oral nutrition generally found in patients undergoing esophageal surgery.

At least 40 kcal per kg of ideal weight should be provided. Therefore, for an ideal patient weighting 70 kg, the recommended caloric intake should be of about 2800 kcal per day.

The protein intake should be included between 1.2 and 1.5 g per kg of ideal weight, provided that no renal disease contraindicates this support.

Proteins should have a high biological value (such as in lean meat, egg white, dried beef, ham, slightly ripened cheeses, ricotta). It could be necessary to decrease the volume of meals using proteins from milk whey or eggs.

These types of food may be combined differently and in such a way as not to increase the total meal volume but just to increase the total amount of proteins.

The intake of carbohydrates must be predominant if there is no concomitant disease that discourages it. The ptyalin digests the simple sugars and the cooked starch, so foods should be cooked for a long time and possibly at high temperatures in order to gelatinize all the starch and make it easier to digest.

In practise, patients should prefer a small quantity of very well-cooked pasta in small quantity or of those types of rice known to be rich in amylopectins (such as thay, arborio, or roma rice) and which should be boiled for a long time. Patients could also eat baked potatoes in small slices, polenta, and slightly toasted oven-baked bread.

Simple sugars, such as sucrose and fructose, can trigger the "dumping syndrome," so they should be limited.

Even fats may trigger the dumping syndrome if they are taken in excessive quantities per meal. The daily fat intake must provide between 20% and 40% of the overall calories amount. This quantity should consist for half of mono-unsaturated fats (such as extra-virgin olive oil) and for the remaining portion of saturated and polyunsaturated fats. The omega-3 fatty acids are the less common in normal diet and their privileged sources are represented by bluefishes and salmon. In general terms, fishes fished in open water should be preferred since farmed fishes have a very low quantity of EPA and DHA.

On the other hand, omega-6 fatty acids are very common in food, and if meat is preferred to fish as a source of animal protein, the omega-6/omega-3 ratio is unbalanced in favor of the former. Among the sources of linolenic acid, they can be noted chia seeds, easily digestible, and linseed oil, despite easily oxidable.

A "soft diet" should be continued for several weeks, without a well defined period, trying to introduce more solid food and slightly larger volumes of oral intake progressively, whenever it is well tolerated.

However, even many months after surgery, the risk of regurgitation is still high, so that chewing food completely should be continued.

Moreover, taking an upright position after any meal and having dinner at least 3 hours before going to bed are good practises not be changed.

Some foods should be avoided immediately after the operation and even in the everyday diet, since they can promote the reflux: coffee, even decaf feinated, tea, alcoholic drinks, and carbonated drinks.

Some spices such as pepper, like chili pepper, and curry should also be avoided, and the same goes for gas-producing vegetables such as cabbages, broccoli, and Brussels sprouts (cruciferous family).

Legumes can be eaten in small quantities. In order to reduce the production of gas, it is advisable to soak vegetables in water for at least 48–72 hours, to change water several times, to cook vegetables in water after making it alkaline by adding bicarbonate, and finally to puree vegetables. The biological value of the amino acids contained in legumes is not high due to the lack of lysine, an essential amino acid. A complementary nutritional intake increasing this biological value and making it similar to meat is representaed by cereals.

Proteins contained in eggs, soy, soft wheat flour, and durum wheat can be added to the purée of legumes in order to achieve a meal with a high biological value of proteic content.

2.7   Summary of the Hygienic-Dietary Recommendations After Esophagectomy

Many institutions have elaborated hygienic-dietary recommendations intended for patients undergoing esophageal resections. Usually, operated patients are suggested:

- To eat soft and moist foods, which are easier to swallow
- To stop eating when they start feeling full
- To wait 1–2 weeks before trying again if a food causes discomfort
- To take small bites and chew food appropriately
- To try eating five to six small meals and snacks during the day rather than three larger meals
- To enrich meals by adding fat, as olive oil, butter, cheese, and cream, in order to increase calories and not volume
- To take the necessary quantity of proteins for muscle mass and immune response
- To eat fish, eggs, cheese, and dairy products if they experience intolerance to meat
- To drink most liquids between the meals (overall 1.5 L/day)
- To sit upright after eating and stay in a sitting or standing position for 45–60 min afterward
- To avoid eating 3 h before bedtime
- To avoid smoking and alcohol
- To check out weight regularly
- To practice physical activity regularly, such as a half-hour slow march every day
- To take nutritional beverage if appetite is poor and weight loss occurs.

These recommendations are surely not exhaustive, but they summarize the most effective attitudes to suggest in order to accelerate and improve patient's recovery.

References

1. De Dominicis F, Jougon J, Brichon P-Y, Tiffet O, Mouroux J, Porte H, et al. Données actuelles sur la chirurgie œsogastrique pratiquée par les équipes de chirurgie thoracique françaises. J Chirurgie Thoracique et Cardio-Vasculaire. 2014;18(2):103–8.
2. Lordick F, Mariette C, Haustermans K, Obermannová R, Arnold D. Oesophageal cancer: ESMO Clinical Practice Guidelines for diagnosis, treatment and follow-up. Ann Oncol. 2016;27(suppl_5):v50–v7.
3. Baker M, Halliday V, Williams RN, Bowrey DJ. A systematic review of the nutritional consequences of esophagectomy. Clin Nutr. 2016;35(5):987–94.
4. Metzger R, Bollschweiler E, Vallböhmer D, Maish M, DeMeester T, Hölscher A. High volume centers for esophagectomy: what is the number needed to achieve low postoperative mortality? Dis Esophagus. 2004;17(4):310–4.
5. Wenner J, Zilling T, Bladström A, Alvegard T, editors. Influence of surgical volume on hospital mortality for cancer of the esophagus and gastric cardia: a national survey in Sweden 1987–96. Proceedings of Book of Abstracts from the European Surgical Association 10th Annual Meeting; 2003.
6. Raymond D. Complications of esophagectomy. Surg Clin North AM. 2012;92(5):1299–313.
7. Blencowe NS, Strong S, McNair AG, Brookes ST, Crosby T, Griffin SM, et al. Reporting of short-term clinical outcomes after esophagectomy: a systematic review. Ann Surg. 2012;255(4):658–66.
8. Poghosyan T, Gaujoux S, Chirica M, Munoz-Bongrand N, Sarfati E, Cattan P. Functional disorders and quality of life after esophagectomy and gastric tube reconstruction for cancer. J Visc Surg. 2011;148(5):e327–e35.

9. Berkelmans GH, van Workum F, Weijs TJ, Nieuwenhuijzen GA, Ruurda JP, Kouwenhoven EA, et al. The feeding route after esophagectomy: a review of literature. J Thorac Dis. 2017;9(Suppl 8):S785.
10. Reim D, Friess H. Feeding challenges in patients with esophageal and gastroesophageal cancers. Gastrointest Tumors. 2015;2(4):166–77.
11. Steenhagen E, van Vulpen JK, van Hillegersberg R, May AM, Siersema PD. Nutrition in peri-operative esophageal cancer management. Expert Rev Gastroenterol Hepatol. 2017;11(7):663–72.
12. Vaezi MF, Pandolfino JE, Vela MF. ACG clinical guideline: diagnosis and management of achalasia. Am J Gastroenterol. 2013;108(8):1238.
13. Schlottmann F, Luckett DJ, Fine J, Shaheen NJ, Patti MG. Laparoscopic Heller myotomy versus peroral endoscopic myotomy (POEM) for achalasia: a systematic review and meta-analysis. Ann Surg. 2018;267:451–60.
14. Zaninotto G, Bennett C, Boeckxstaens G, Costantini M, Ferguson M, Pandolfino J, et al. The 2018 ISDE achalasia guidelines. Dis Esophagus. 2018;31(9):doy071.
15. Richards WO, Torquati A, Holzman MD, Khaitan L, Byrne D, Lutfi R, et al. Heller myotomy versus Heller myotomy with Dor fundoplication for achalasia: a prospective randomized double-blind clinical trial. Ann Surg. 2004;240(3):405.
16. Petrucciani N, de'Angelis N, Brunetti F. Robotic Toupet fundoplication following Heller myotomy for achalasia (with video). J Visc Surg. 2018;155:427–8.
17. Rebecchi F, Giaccone C, Farinella E, Campaci R, Morino M. Randomized controlled trial of laparoscopic Heller myotomy plus Dor fundoplication versus Nissen fundoplication for achalasia: long-term results. Ann Surg. 2008;248(6):1023–30.
18. Campos GM, Vittinghoff E, Rabl C, Takata M, Gadenstätter M, Lin F, et al. Endoscopic and surgical treatments for achalasia: a systematic review and meta-analysis. Ann Surg. 2009;249:45–57.
19. Weimann A, Braga M, Carli F, Higashiguchi T, Hübner M, Klek S, et al. ESPEN guideline: clinical nutrition in surgery. Clin Nutr. 2017;36(3):623–50.
20. Cederholm T, Barazzoni R, Austin P, Ballmer P, Biolo G, Bischoff SC, et al. ESPEN guidelines on definitions and terminology of clinical nutrition. Clin Nutr. 2017;36(1):49–64.
21. Bower MR, Martin RC. Nutritional management during neoadjuvant therapy for esophageal cancer. J Surg Oncol. 2009;100(1):82–7.
22. Haverkort E, Binnekade J, Busch O, van Berge Henegouwen M, De Haan R, Gouma D. Presence and persistence of nutrition-related symptoms during the first year following esophagectomy with gastric tube reconstruction in clinically disease-free patients. World J Surg. 2010;34(12):2844–52.
23. Greene CL, DeMeester SR, Worrell SG, Oh DS, Hagen JA, DeMeester TR. Alimentary satisfaction, gastrointestinal symptoms, and quality of life 10 or more years after esophagectomy with gastric pull-up. J Thorac Cardiovasc Surg. 2014;147(3):909–14.
24. Donohoe CL, McGillycuddy E, Reynolds JV. Long-term health-related quality of life for disease-free esophageal cancer patients. World J Surg. 2011;35(8):1853–60.
25. Martin L, Lagergren P. Long-term weight change after oesophageal cancer surgery. Br J Surg. 2009;96(11):1308–14.
26. Ouattara M, D'Journo XB, Loundou A, Trousse D, Dahan L, Doddoli C, et al. Body mass index kinetics and risk factors of malnutrition one year after radical oesophagectomy for cancer. Eur J Cardiothorac Surg. 2012;41(5):1088–93.
27. Paik C, Choi MG, Lim C, Park J, Chung W, Lee KM, et al. The role of small intestinal bacterial overgrowth in postgastrectomy patients. Neurogastroenterol Motil. 2011;23(5):e191–e6.
28. Al-Hadrani A, Lavelle-Jones M, Kennedy N, Neill G, Sutton D, Cuschieri A, editors. Bile acid malabsorption in patients with post-vagotomy diarrhoea. Ann Chir Gynaecol. 1992;81:351–3.
29. Huddy JR, Macharg FM, Lawn AM, Preston SR. Exocrine pancreatic insufficiency following esophagectomy. Dis Esophagus. 2013;26(6):594–7.
30. Heneghan HM, Zaborowski A, Fanning M, McHugh A, Doyle S, Moore J, et al. Prospective study of malabsorption and malnutrition after esophageal and gastric cancer surgery. Ann Surg. 2015;262(5):803–8.

31. Haverkort EB, Binnekade JM, de Haan RJ, Busch OR, van Berge Henegouwen MI, Gouma DJ. Suboptimal intake of nutrients after esophagectomy with gastric tube reconstruction. J Acad Nutr Diet. 2012;112(7):1080–7.
32. Ryan AM, Rowley SP, Healy LA, Flood PM, Ravi N, Reynolds JV. Post-oesophagectomy early enteral nutrition via a needle catheter jejunostomy: 8-year experience at a specialist unit. Clin Nutr. 2006;25(3):386–93.
33. National oesophago-gastric cancer audit. Royal College of Surgeons of England; 2012.
34. Zheng Y-M, Li F, Qi B-J, Luo B, Sun H-C, Liu S, et al. Application of perioperative immunonutrition for gastrointestinal surgery: a meta-analysis of randomized controlled trials. Asia Pac J Clin Nutr. 2007;16(S1):253–7.
35. Montejo JC, Zarazaga A, López-Martínez J, Urrútia G, Roqué M, Blesa AL, et al. Immunonutrition in the intensive care unit. A systematic review and consensus statement. Clin Nutr. 2003;22(3):221–33.
36. Heyland DK, Novak F, Drover JW, Jain M, Su X, Suchner U. Should immunonutrition become routine in critically ill patients? A systematic review of the evidence. JAMA. 2001;286(8):944–53.
37. Beale RJ, Bryg DJ, Bihari DJ. Immunonutrition in the critically ill: a systematic review of clinical outcome. Crit Care Med. 1999;27(12):2799–805.
38. Heys SD, Walker LG, Smith I, Eremin O. Enteral nutritional supplementation with key nutrients in patients with critical illness and cancer: a meta-analysis of randomized controlled clinical trials. Ann Surg. 1999;229(4):467.
39. Patel DA, Naik R, Slaughter JC, Higginbotham T, Silver H, Vaezi MF. Weight loss in achalasia is determined by its phenotype. Dis Esophagus. 2018;31 https://doi.org/10.1093/dote/doy046.
40. Patel DA, Lappas BM, Vaezi MF. An overview of achalasia and its subtypes. Gastroenterol Hepatol. 2017;13(7):411.
41. Newberry CA, Vajravelu RK, Lynch KL. Obese achalasia patients are at significant nutritional risk. Gastroenterology. 2017;152(5):S197.
42. Zifodya JS, Kim HP, Silver HJ, Slaughter JC, Higginbotham T, Vaezi MF. Tu1199 nutritional status of patients with untreated achalasia. Gastroenterology. 2015;148(4):S-819–S-20.
43. Rohof W, Bredenoord A. Chicago classification of esophageal motility disorders: lessons learned. Curr Gastroenterol Rep. 2017;19(8):37.
44. Furuzawa-Carballeda J, Aguilar-León D, Gamboa-Domínguez A, Valdovinos M, Nuñez-Álvarez C, Martín-del-Campo L, et al. Achalasia—an autoimmune inflammatory disease: a cross-sectional study. J Immunol Res. 2015;2015:729217.
45. Hungness ES, Teitelbaum EN, Santos BF, Arafat FO, Pandolfino JE, Kahrilas PJ, et al. Comparison of perioperative outcomes between peroral esophageal myotomy (POEM) and laparoscopic Heller myotomy. J Gastrointest Surg. 2013;17(2):228–35.
46. Salvador R, Pesenti E, Gobbi L, Capovilla G, Spadotto L, Voltarel G, et al. Postoperative gastroesophageal reflux after laparoscopic Heller-Dor for achalasia: true incidence with an objective evaluation. J Gastrointest Surg. 2017;21(1):17–22.
47. Finan KR, Renton D, Vick CC, Hawn MT. Prevention of post-operative leak following laparoscopic Heller myotomy. J Gastrointest Surg. 2009;13(2):200.
48. Leeds SG, Burdick J, Ogola GO, Ontiveros E, editors. Comparison of outcomes of laparoscopic Heller myotomy versus per-oral endoscopic myotomy for management of achalasia. Proc (Bayl Univ Med Cent). 2017;30:419–23.
49. Network UH. University Health Network (UHN). Patient education. Heller Myotomy. 2016.
50. Rosemurgy A, Villadolid D, Thometz D, Kalipersad C, Rakita S, Albrink M, et al. Laparoscopic Heller myotomy provides durable relief from achalasia and salvages failures after botox or dilation. Ann Surg. 2005;241(5):725.
51. Frantzides CT, Carlson MA. Atlas of minimally invasive surgery. Philadelphia, PA: Saunders; 2009.

Nutritional Support After Surgery of the Stomach

3

Donato Francesco Altomare and Patrizia Ancona

3.1 Introduction

Total or partial resection of the stomach is a major surgery performed mainly for complications of peptic ulcers (bleeding, stricture, perforation) or for gastric cancer. Other less frequent indications are benign tumors and severe gastroparesis. Gastric surgery for obesity will be treated in a separate chapter.

While surgery for peptic ulcers has dramatically decreased because of the large use of proton pump inhibitors, effective antibiotic therapy against *Helicobacter pylori*, and technical improvements in endoscopy, gastric cancer remains the most frequent indication for gastric resection. Gastric cancer is, in fact, the second/third most frequent cancer of the gastrointestinal tract with a poor prognosis in the advanced stages. Environmental, genetic, and chronic gastritis with *Helicobacter pylori* contribute to the development of gastric cancer. Surgical resection with regional lymph node dissection by open or laparoscopic surgery may be potentially curative only at early stages which are difficult to diagnose because of the paucity of specific symptoms.

The extent of the gastric resection is conditioned by the preoperative stage and tumor location since subtotal gastrectomy is the standard for antropyloric cancers and total gastrectomy for those invading the other parts of the stomach and the gastroesophageal junction. In case of peptic ulcer, a more limited resection of the stomach is required, and the most frequent modalities of reconstruction include the Billroth I and II and more rarely the Roux-en-Y gastrojejunal bypass.

All these operations determine important changes in the pathophysiology of nutrition because of the important role of the stomach in digestion of foods and iron metabolism and because of the changes in the GI motility and transit.

D. F. Altomare (✉)
Department of Emergency and Organ Transplantation, University "Aldo Moro" of Bari, Bari, Italy
e-mail: donatofrancesco.altomare@uniba.it

P. Ancona
Nutritionist Via Gramsci, Bari, Italy

© Springer Nature Switzerland AG 2019
D. F. Altomare, M. T. Rotelli (eds.), *Nutritional Support after Gastrointestinal Surgery*, https://doi.org/10.1007/978-3-030-16554-3_3

3.2 Role of the Stomach in the Physiology of Nutrition

The stomach plays a pivotal role in the physiology of nutrition. Due to its dimension and compliance, it is able to hold up to 2 L of food and liquid.

When the ingested food enters the stomach, its contraction and secretions are activated to mix the food with gastric juice to produce the chyme.

Gastric secretions: The gastric juice is composed by hydrochloric acid and mucus. The hydrochloric acid is actively secreted into the stomach by the parietal cells of the mucosa by a complex and energy-requiring mechanism in response to various central, hormonal (gastrin, VIP, somatostatin, glucagon, GIP (gastric inhibitory peptide or enterogastrone), substance P, etc.), and local stimuli, keeping the gastric pH around 1.5–3. Such an acid environment allows conversion of the pepsinogen into the active enzyme pepsin and causes denaturation of the food proteins, with exposure of sites of these proteins where the pepsin can break chemical ligands thus generating peptides. Furthermore, the strong acid environment of the stomach can kill the majority of the pathogen bacteria ingested with the foods. The mucus, on the other hand, plays a major role against the damage of the mucosa cells by the hydrochloric acid.

The intrinsic factor (IF) is a glycoprotein secreted by the parietal cells of the gastric mucosa, which is able to link the vitamin B12 in order to allow its absorption in the ileum where specific receptors for the IF are present. Its absence after gastrectomy is responsible for a severe anemia due to vitamin B12 deficiency.

Other components of the gastric juice are acetylcholine, histamine, and gastrin secreted by the post-gangliar vagal nerves, mast cells, and G cells of the antrum, respectively, and involved in various ways in the regulation of the hydrochloric acid and pepsinogen secretion, gastric vascular supply, motility, and trophism.

Gastric motility: Gastric motility is aimed to mix the food with gastric juice to form the chyme and to promote the passage of the chyme through the pylorus into the duodenum. It is a very fine and complex mechanism controlled by the extrinsic parasympathetic innervation of the vagus nerve, by the intrinsic gastric pacemaker usually localized on the great curvature near to the fundus and the upperpart of the gastric body and by several peptides secreted by specialized cells into the duodenum and jejunum like the GIP, motilin, and somatostatin.

3.3 Short Description of the Main Surgical Operations on the Stomach

While the demolitive phase of this operation is well standardized, the modality of reconstruction of the GI tract continuity has dozens of modalities proposed. In this chapter, for brevity we will consider only the most frequent ones, the omega gastrojejunal bypass according to Hofmeister-Finsterer after subtotal gastrectomy which for some pathophysiologic consequences resembles the Billroth II technique, the gastroduodenal reconstruction according to Billroth I (now rarely performed) and Roux-Orr Y esophageal-jejuno anastomosis after total gastrectomy.

In all these operations, the lesser omentum is divided and with-it part or all the hepatic branches of the anterior vagus nerve, and in case of total gastrectomy, both vagus nerves are interrupted leaving all the intra-abdominal viscera (except from the pelvic organs) parasympathetically denervated.

Hypotonia of the biliary tract and gallbladder, in these cases, could favor stone formation.

3.3.1   Subtotal Gastrectomy with Omega Gastrojejunostomy

In this operation, the stomach is fully mobilized, the right gastric and gastroepiploic arteries are ligated and interrupted, and the second part of the duodenum is mobilized (Kocher maneuver). The duodenum is closed and interrupted just after the pylorus, and the left gastric artery is ligated and interrupted just after the origin of the ascending branch to the gastric fundus. The stomach is then resected leaving the fundus and, if possible, part of the gastric body. In case of gastric cancer, the macroscopic proximal margin of clearance should be about 5 cm, and a complete lymph nodes dissection is mandatory to achieve a curative operation. In the Hofmeister-Finsterer reconstruction, the gastric remnant is anastomosed antecolic with a jejunal loop 50–60 cm far from the ligament of Treitz (Fig. 3.1). The efferent part of the jejunal loop must be on the greater curvature side. The afferent loop should be short, or an entero-entero anastomosis according to Braun is suggested at the foot of the omega loop in order to prevent the passage of the bile and pancreatic juice through

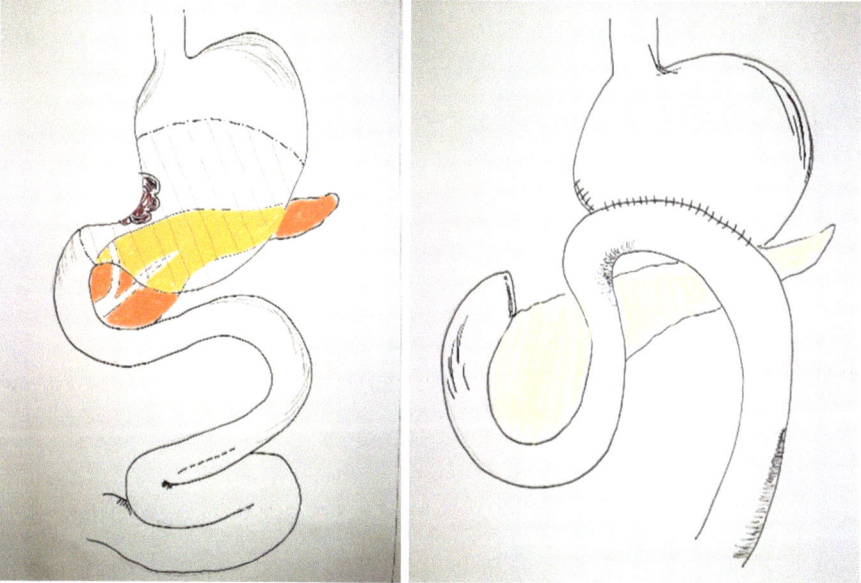

Fig. 3.1   Partial gastrectomy with gastrojejunostomy (Billroth II)

the residual stomach. The gastrojejunal anastomosis can involve the entire residual stomach or, better, only part of it on the side of the greater curvature.

This type of gastrointestinal reconstruction was frequently performed in the last century with the main indication of antropyloric peptic ulcer despite the first case was performed by Theodor Billroth for pylorus cancer in 1881. The extent of the gastric resection depends on the indication (cancer or peptic ulcer) and the location of the ulcer. In benign cases about 2/3 of the stomach is resected in order to eliminate the HCl secretory part of the mucosa and prevent the recurrence of the ulcer. After mobilization of the gastrocolic ligament (or the greater omentum in case of cancer) and the lesser omentum, the right gastric and the gastroepiploic arteries are ligated and the proximal duodenum mobilized and divided close to the pylorus. A vagotomy is also usually performed during this operation. The resection of the stomach is performed following an imaginary line close to the cardias on the lesser curvature and on the mid-greater curvature leaving about 40–50% of the stomach. The upper part of the stomach is closed, and an end-to-end gastroduodenal anastomosis is performed by hand suture (Fig. 3.2).

The total removal of the stomach is performed for cancer arising in the cardias and in the proximal 2/3 of the stomach. In this operation the mobilization of the stomach already described in the partial gastrectomy is completed by the section of the main trunk of the left gastric artery, the left gastroepiploic artery, and the short

Fig. 3.2 Partial gastrectomy with gastroduodenostomy (Billroth I)

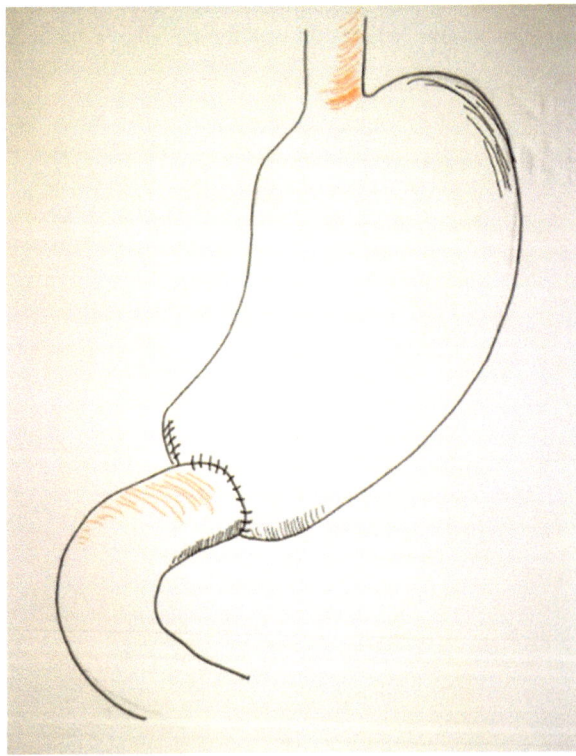

Fig. 3.3 Roux-Orr en Y *oesophago-jejunostomy* reconstruction following total gastrectomy

gastric arteries. The jejunum is divided about 60–70 cm distal to the ligament of Treitz, and its distal end is then anastomosed with the esophagus end to end (hand suture) or end to side (stapler suture according to Orr). The proximal part of the jejunum containing the bile and entero-pancreatic juice, is anastomosed to the distal jejunum about 40–50 cm from the esophageal anastomosis thus creating a Y (Roux en Y) (Fig. 3.3). Several variations of this operation have been described; however this is the most common way of reconstruction after total gastrectomy.

3.4 Pathophysiology of the GI Tract After Total or Subtotal Gastrectomy

Billroth I operation was originally preferred because of the lesser distortion of the gastrointestinal anatomy, since the chyme meet bile, pancreatic, and duodenal secretions directly after leaving the remnant stomach, whereas in case of gastrojejunostomy, the duodenum is functionally excluded, and the chyme mix with bile and pancreatic juice only in the mid-jejunum. On the other hand, in both methods the passage of the food into the intestine is rapid and uncontrolled due to the absence of the antropyloric region and the damage to the extrinsic parasympathetic innervation. Lack of coordination of the gallbladder bile and pancreatic secretion into the duodenum is an obvious consequence of the gastrointestinal bypass.

In total gastrectomy, the physiology of nutrition is profoundly disturbed because there is no barrier to jejunum-esophageal reflux (no cardias), there is no gastric reservoir, the food will never have contact with hydrochloric acid and pepsin, the mixing of this undigested food with bile and pancreatic juice occurs only after several centimeters of jejunum, both vagus nerves have been interrupted, and the pacemaker area of the stomach is absent.

The following GI disturbances can follow these operations:

3.4.1 Dumping Syndrome (DS)

According to Moore's definition, the dumping syndrome is the "occurrence of sweating, unpleasant warmth, flushing, nausea, epigastric fullness, palpitation, or explosive diarrhea during or within 15 min after eating, and lasting up to 45 min, and partially or completely relieved by lying down". This syndrome complicates about 20–50% of the Billroth II-type gastrojejunostomy although has been described also in Billroth I reconstruction. It can be however clinically severe in only 1–5% of the cases.

The pathogenesis of the DS is due to the rapid transit of foods from the residual stomach to the intestine due to the absence of the antropyloric region. In fact the capacity of the residual stomach is strongly reduced, and its accommodation is impaired because of the parasympathetic denervation for the vagotomy. The pylorus is removed, and therefore the ingested food is rapidly reversed into the jejunum, bypassing the duodenum which cannot trigger its feedback inhibition on the gastric motility. Also the gastric acid and enzymes are drastically reduced.

Such a rapid passage of hyperosmolar foods into the small bowel favors the rapid release of serotonin responsible for mesenteric and peripheral vasodilation, causing hypotension.

Other gut hormones such as glucagon-like peptide 1 (GLP-1), vasoactive intestinal polypeptide (VIP), enteroglucagon, peptide YY, and pancreatic polypeptide, which are released after food ingestion, have higher serum concentrations in patients with DS and probably play a role in the onset of dumping syndrome [1–3].

3.4.2 Late Dumping

A further metabolic alteration related to the rapid delivery of foods with high concentration of carbohydrates has been described after gastrectomy and is known as late dumping syndrome. It occurs about 1–3 h after a meal and is characterized by malaise, hypotension, cold sweating, and hypoglycemia, probably due to a high serum insulin levels as a response to the rapid hyperglycemia after meal.

3.4.3 Roux-en-Y Stasis Syndrome

The motility of the small bowel after gastrectomy is disturbed by the interruption of the vagus nerve and the altered regulation of the enteric hormones. Furthermore, in Roux-en-Y total gastrectomy, there is a displacement of the high-frequency intestinal pacemakers leading to retrograde peristaltic activities or severe dysmotility of the afferent limb. In normal physiology, the GI pacemakers are set in a craniocaudal disposition according to their frequency of firing, the higher-frequency pacemaker being in the greater curvature of the stomach. In the Roux-en-Y reconstruction, the jejunal loop anastomosed to the esophagus has a lower-frequency pacemaker compared to the proximal jejunal loop of the entero-enteric anastomosis [4–6].

As a result of the absence of peristaltic movements or even antiperistaltic contraction, the passage of the food can be impaired leading to a dilatation of the intestinal loop.

3.4.4 Blind Loop Syndrome

The afferent jejunal loop in Billroth II-type reconstruction can present motility dysfunctions and accumulation of wasting organic materials leading to the blind loop syndrome [7], a pathologic condition with dilatation and stasis of the intestinal loop which allows abnormal bacterial overgrowth, bile salt breakdown resulting in diarrhea, malnutrition, and weight loss.

3.4.5 Biliary Lithiasis

Formation of stones in the gallbladder and sometimes in the extrahepatic biliary tract is common in these patients, due to their hypotonia because of the parasympathetic denervation and because of the lack of the coordinated biliary secretion which follows the normal passage of the chyme into the duodenum. For that reason, a prophylactic removal of the gallbladder is often performed, especially after total gastrectomy.

3.4.6 Nutritional Deficiencies

Several nutritional deficiencies usually develop after total or subtotal gastrectomy, in particular pernicious anemia, due to the absence or strong reduction of the cyanocobalamin (vitamin B12) absorption which is essential for promoting purine and thymidylate syntheses and DNA synthesis in cells with rapid turnover like erythrocytes. Its deficiency associated to folate and iron deficiency causes erythroblast apoptosis, leading to anemia from ineffective erythropoiesis. Impairment of iron absorption into the duodenum is another cause of iron deficiency anemia often present in these patients [8].

Malabsorption with steatorrhea and diarrhea may result from incomplete digestion of proteins and lipids because of the absence of pepsin and hydrochloric acid and late contact of the ingested food with the pancreatic juice. Clinical manifestations include severe dyspepsia, meteorism, diarrhea, and weight loss.

3.5   Nutritional Support After Gastrectomy for Gastric Cancer

Gastrectomy is the treatment of choice in gastric cancer, and the majority of gastric cancer patients, after surgery, present malnutrition phenomena which leads to reduced responses to the treatment [9] and increases treatment-associated side effects [10] often inducing changes in the treatment protocol and sometimes until to determine treatment interruption. Partial or full gastrectomy reduces the food intake, while the effects of vagal interruption can cause abdominal distension, discomfort, and frequent bowel movement [11]. The patient who has undergone partial or full gastrectomy must adhere to a targeted diet, to specific rules to be followed for proper nutrition, to continuous medical checks and, if necessary, to a chemoradiotherapy. Chemotherapy following gastrectomy also induces anorexia, sore throat, dry mouth, taste change, nausea, diarrhea, constipation, and fatigue which eventually contribute to weight loss and malnutrition [12]. High risk of malnutrition among gastrectomy patients was shown to delay the rate of recovery and increase cancer-related deaths [13].

Nutritional support in gastrectomy patient, therefore, is very important to:

- Prevent weight loss
- Promote tissue healing after surgery
- Integrate nutritional deficiencies caused by stomach removal
- Enhance the immune system
- Eliminate or reduce the symptomatology caused by the different postgastrectomy syndromes

After partial or total removal of the stomach, the food chewed in the mouth passes more quickly, or directly, to the first part of the intestine. Therefore, the phase of digestion that was first carried out in the stomach through gastric juices is reduced or missing [14]. Postgastrectomy symptoms and disorders can be alleviated with some changes on nutritional support and lifestyle. There are no general rules valid for everyone, but each person will adjust himself by listening to the signals of his body.

Adequate postoperative support can help minimize early or late depletion syndrome, sense of early satiety, diarrhea, fat malabsorption, malabsorption of lactose, anemia, weight loss, and reflux [15].

In the days following the surgery, the patient may have digestive problems as a result of a new anatomical and functional condition. Therefore, it is useful to make minimum nutritional shares at least six times a day, favoring foods with higher nutritional value, and then to increase them as a result of personal tolerance every 3–4 days [16, 17].

Food to be preferred, but with moderation:

- Complex carbohydrates (for the slow release of sugars and greater control of blood sugar): pasta, rice, and bread, preferring pasta with fresh tomatoes or vegetables.

- Meat, discarding visible fat and poultry skin.
- Fresh or frozen fish.
- Cooked/raw ham without fat and bresaola.
- Fresh vegetables cooked and preferably passed.
- Milk, preferably not skimmed, in small quantities. If not tolerated, use high digestibility milk.
- Yogurt: take at least a portion a day, preferring the one from natural not skimmed milk.
- Extra-virgin olive oil.
- Butter.

3.5.1   Dumping Syndrome

It occurs when the bolus is rapidly introduced into the small intestine triggering gastrointestinal and/or vasomotor symptoms. The following nutritional advices are valid for both the late and early dumping syndromes.

3.5.1.1 Nutritional Advices for Preventing the Dumping Syndrome
- Avoid taking fizzy drinks and alcohol.
- Reduce the intake of white sugar, raw or in the form of syrup.
- Reduce the intake of foods or drinks that have as their first ingredients sugar, sorbitol, and xylitol (e.g. biscuits, various sweets, candies, sugary drinks), and prefer bread, crackers, rusks, pasta, and rice.
- Use type I and type II flours instead of flour 00 or 0.
- Eat fresh fruits (apples, pears, peaches, plums, citrus fruits, bananas), carrots, and potatoes.
- Avoid nuts without shells (raisins, prunes, apricots, etc.) and sweetened fruit juices.
- Limit the amount of raw green leafy vegetables, cabbage, broccoli, cauliflower, corn, turnips, and legumes if they give an excessive sense of swelling or satiety. Do not combine them in the same meal.
- Prefer cooked vegetables without seeds and peel, filtrated and cetrifuged vegetables.
- Use caloric snacks as needed, like nuts and yogurt without added sugar (if you are not lactose intolerant).
- Take protein-rich foods regularly: fish (avoiding fat such as eel, mullet, lobster, and fish preserved in oil), lean meat (preferably soft and well cooked), cheese (if you are not lactose intolerant), eggs, and legumes. Because the legumes can cause swelling, prefer the de-hulled ones.

Dumping syndrome unresponsive to diet manipulation may require meeting with a nutritionist and use of gut-slowing medication [18].

3.5.2 Sense of Early Satiety or Gastric Stasis

It is the feeling of satiety which occurs after eating only small amounts of food. It is linked to the difficulty of the intestine to expand to accommodate the food. Postprandial bloating, discomfort, or fullness lasting many hours are specific symptoms of gastric stasis. Emesis of undigested food ingested hours to days before may also be present. These patients can develop bezoar formation, bacterial overgrowth, and intolerance to solid food; liquids may be processed normally or rapidly [14, 19].

3.5.2.1 Nutritional Support for the Sense of Early Satiety
- Make small and frequent meals, if possible at the same times to give a "rhythm" to the body.
- Eat while sitting, chew your food well before swallowing, and do not go to bed as soon as you eat (stay up at least a couple of hours).
- Wear clothes that do not tighten at the waist.
- Initially avoid bran and foods rich in fiber (pasta, rice and wholemeal bread, legumes with peel). It will probably be possible to introduce them later but gradually and do not take them together in one meal.
- Drink little during meals. Avoid carbonated water and carbonated soft drinks.
- Drink preferably away from meals.
- Try to arrange the food in large dishes, not to have the feeling of eating a lot.

3.5.3 Diarrhea

If it occurs 15–30 min after the meal, a dumping syndrome's symptom should be considered.

3.5.3.1 Nutritional Support to Control Diarrhea Post Gastrectomy
1. Drink often in small sips between meals to prevent dehydration.
2. Avoid drinks that contain alcohol or caffeine.
3. Avoid consumption of whole grains, green leafy vegetables, flax seeds, bran, and products added with bran.
4. Prefer rice, semolina, and millet.
5. Prefer citrus fruits, fresh fruits (apples, pears, banana), carrots, and potatoes.
6. Avoid foods that can make sense of garlic, onion, broccoli, sweets, cheeses and dairy products, and legumes with peel.
7. Avoid foods that are too hot or too cold.

3.5.4 Fat Malabsorption

Fat absorption may decrease after partial or total removal of the stomach. Fats in the stool may increase and "fat" feces may be noted. It can also decrease the absorption of some vitamins (vitamin A, vitamin D, vitamin K, and vitamin E) and calcium [14].

Fat absorption may decrease because of:

- Lack or decrease of digestive enzymes
- Less contact between ingested food, digestive enzymes, and biliary salts

3.5.4.1 Nutritional Support to Help for Fat Malabsorption
- Add the fats gradually. At the beginning, use fats only in small quantities. Then increase their use, making sure to tolerate them well.
- Among the fat foods, prefer extra-virgin olive oil, fish, nuts, oil seeds, and avocado.
- Limit fried foods, cream, mascarpone, bacon, and lard. It is possible indicate an integration of digestive enzymes to aid digestion and vitamins (A, D, E, K) and calcium.

3.5.5 Malabsorption of Lactose

After the partial or total removal of the stomach, some people may develop malabsorption of lactose presenting swollen or painful tummy, meteorism, cramps, or diarrhea after eating dairy products. When these symptoms occur, a lactose breath test should be prescribed to check for malabsorption [20].

3.5.5.1 Nutritional Support in Case of Lactose Malabsorption
- Avoid cow's milk and fresh cheeses made with this milk (e.g., mozzarella, stracchino, flakes of milk, etc.), spreadable cheeses, cow's ricotta, cream and béchamel (not vegetables), sweets with creams, and ice creams made from milk or cream vaccines.
- If tolerated, prefer mature cheeses (such as Parmesan or other cheeses with more than 6 months) and cheeses and ricotta of goat or sheep for their low lactose content.
- Use milk without lactose or vegetable, such as oats, rice, almond, and soy.
- Use products prepared with lactose-free milk or vegetable milk. It is possible to indicate an integration (even temporary) of enzymes that can facilitate lactose digestion.

3.5.6 Anemia

Anemia is a condition that almost always arises after partial or total removal of the stomach due to iron, vitamin B12, and folic acid (vitamin B9) deficiencies. The patients' blood levels should be monitored regularly [21].

3.5.6.1 Nutritional Support to Prevent Postgastrectomy Anemia
Take iron-rich foods with foods rich in vitamin C, which help absorb it. An example can be lemon juice (rich in vitamin C) squeezed on foods rich in iron (e.g., radicchio and meat).

Foods rich in iron are beef, horse, lamb, goat, pork, liver (not processed), mollusks (esp. clams), sardines, tuna (not processed), shrimps, eggs, lentils, spinach, pumpkin seeds (not salted or toasted), dried fruit, and dark chocolate. In any case it is advisable not to exceed with the consumption of these foods. Some of them, in fact, are rich in cholesterol (e.g., red meat, eggs, liver) or rich in fiber (e.g., spinach).

Food rich in vitamin C are tomatoes, potatoes (without peel), citrus fruits, pineapples, strawberries, and raspberries. Consume the freshly cut fruit and its freshly prepared juices because vitamin C is sensitive to air and heat.

Foods rich in folic acid are liver, offal, spinach, chard, tomatoes, oranges, dried fruit, oilseeds, and green leafy vegetables. Consume these foods either raw or steamed, to limit the loss of folic acid.

Follow any supplement or therapy indicated by the doctor.

3.5.7 Weight Loss

In the first months after the partial or total gastrectomy, weight loss can occur. Then the weight should return to normal values. It is very important to keep your weight under control and to tell your doctor if there are important weight drops in a short time.

3.5.7.1 Nutritional Support to Gain Body Weight
- Make calories with nuts, dried fruit, yogurt (without added sugar), or pieces of cheese (if lactose is tolerated).
- In case of lack of appetite, prefer foods that bring a good variety of nutrients (e.g., fish and boiled potatoes, cereals and legumes, etc.).
- Use foods that are easy to digest. Yogurt, for example, is more digestible than milk.

3.5.8 Reflux

After subtotal or total gastrectomy, an esophageal reflux is very likely to occur, particularly when the cardias have been removed. The most common symptom of reflux is a burning sensation, perceived in the central part of the thorax. Reflux is generally more common at night or when lying down.

3.5.8.1 Nutritional Support to Limit Esophageal Reflux
- Make small and frequent meals.
- Avoid drinking before bedtime.
- Avoid stretching after a meal or eating from a bed.
- Limit foods or drinks such as chocolate, carbonated drinks, caffeine or alcohol, fried foods, and overly spicy foods.

3.6   Postgastrectomy Example Diet

Breakfast

- 1 plain yogurt
- 1 slice whole wheat toast
- 1/2 banana
- 1 herbal tea (drink 30–60 min after eating).

Midmorning

- 40 g Parmesan cheese
- Fresh fruit (150 g)

Lunch

- 150 g hamburger with lettuce and tomato
- 1 small bun
- 1/2 cup unsweetened fruit cocktail
- 1/2 cup vegetable juice (drink 30–60 min after eating)

Midafternoon

- 1 tablespoon peanut butter
- 1 slice whole wheat toast
- 1/2 cup orange juice (drink 30–60 min after eating)

Dinner

- Steamed fish
- 150 g of mashed potatoes
- 200 g of green beans
- 3 teaspoons of extra-virgin olive oil (evo)
- 1 cooked apple
- 1 herbal tea (drink 30–60 min after eating)

After dinner

- 10 peeled almonds
- 1 herbal tea (drink 30–60 min after eating)

References

1. Moore HG Jr. Complications of gastric surgery. In: Harkins HN, Nyhus LM, editors. Surgery of the stomach and duodenum. 2nd ed. London: J&A Churchill Ltd; 1969. p. 683.
2. Davis JL, Ripley RT. Postgastrectomy syndromes and nutritional considerations following gastric surgery. Surg Clin North Am. 2017;97(2):277–93.
3. Mala T, Hewitt S, Høgestøl IK, Kjellevold K, Kristinsson JA, Risstad H. Dumping syndrome following gastric surgery. Tidsskr Nor Laegeforen. 2015;135(2):137–41.
4. Gustavsson S, Ilstrup DM, Morrison P, Kelly KA. Roux-Y stasis syndrome after gastrectomy. Am J Surg. 1988;155(3):490–4.
5. Bassotti G, Gullà P, Betti C, Whitehead WE, Morelli A. Manometric evaluation of jejunal limb after total gastrectomy and Roux-Orr anastomosis for gastric cancer. Br J Surg. 1990;77(9):1025–9.
6. Altomare DF, Rubini D, Pilot MA, Farese S, Rubini G, Rinaldi M, Memeo V, D'Addabbo A. Oral erythromycin improves gastrointestinal motility and transit after subtotal but not total gastrectomy for cancer. Br J Surg. 1997;84(7):1017–21.
7. Starlz TE, Butz GW Jr, Hartman CF. The blind-loop syndrome after gastric operations. Surgery. 1961;50:849–58.
8. Ruz M, Carrasco F, Rojas P, Codoceo J, Inostroza J, et al. Iron absorption and iron status are reduced after Roux-en-Y gastric bypass. Am J Clin Nutr. 2009;90:527–32.
9. Salas S, Deville JL, Giorgi R, Pignon T, Bagarry D, Barrau K, Zanaret M, Giovanni A, Bourgeois A, Favre R, et al. Nutritional factors as predictors of response to radio-chemotherapy and survival in unresectable squamous head and neck carcinoma. Radiother Oncol. 2008;87(2):195–200.
10. Cessot A, Hebuterne X, Coriat R, Durand JP, Mir O, Mateus C, Cacheux W, Lemarie E, Michallet M, de Montreuil CB, et al. Defining the clinical condition of cancer patients: it is time to switch from performance status to nutritional status. Support Care Cancer. 2011;19(7):869–70.
11. Seung H, Sung-Eun K, Yoon-Koo K, Baek-Yeol R, Min-Hee R, et al. Association of nutritional status-related indices and chemotherapy-induced adverse events in gastric cancer patients. BMC Cancer. 2016;16:900.
12. Grant B, Byron J. Nutritional implications of chemotherapy. In: Elliot L, Molseed LL, McCallum PD, editors. The clinical guide to oncology nutrition. 2nd ed. Chicago: American Dietetic Association; 2006.
13. Santarpia L, Contaldo F, Pasanisi F. Nutritional screening and early treatment of malnutrition in cancer patients. J Cachex Sarcopenia Muscle. 2011;2(1):27–35.
14. Amy E. Radigan post-gastrectomy: managing the nutrition fall-out. Pract Gastroenterol. 2004;28(6):63–75. Nutrition Issues in Gastroenterology, series 18. Carol Rees Parrish, R.D., MS, Series Editor
15. Rogers C. Postgastrectomy nutrition. Nutr Clin Pract. 2011;26(2):126–36.
16. Liedman B, Svedlund J, Sullivan M, et al. Symptom control may improve food intake, body composition, and aspect of quality of life after gastrectomy in cancer patients. Dig Dis Sci. 2001;46:2673–80.
17. Gustafsson UO, et al. Guidelines for preoperative care in elective colonic surgery: Enhanced Recovery After Surgery (ERAS®) Society recommendations. Clin Nutr. 2012;31:783–800.
18. Harju E. Metabolic problems after gastric surgery. Int Surg. 1990;75:27–35.
19. Meyer J. Chronic morbidity after ulcer surgery. In: Sleisenger M, Fordtran J, editors. Gastrointestinal diseases. 5th ed. Philadelphia: Saunders; 1994. p. 731–44.
20. McCray S. Lactose intolerance: considerations for the clinician. Pract Gastroenterol. 2003;Series 2:21–39.
21. Adachi S, Kawamoto T, Otsuka M, et al. Enteral vitamin B12 supplements reverse postgastrectomy B12 deficiency. Ann Surg. 2000;232:199–201.

Nutritional Support After Cholecystectomy

4

Donato Francesco Altomare and Maria Teresa Rotelli

4.1 Cholecystectomy

4.1.1 Indications

Cholecystectomy for gallstone is one of the most frequently performed abdominal operations because of the large diffusion of the gallstone disease in the western population and for the adoption in the last 25 years of the laparoscopic cholecystectomy technique instead of the open surgery [1]. In fact, the minimal invasiveness of the laparoscopic procedure has extended the indication for gallbladder removal even if the international guidelines discourage the operation in cases of asymptomatic gallstones accidentally discovered during an abdominal ultrasound.

Cholecystectomy is also performed for asymptomatic adenomas growing into the gallbladder and, of course, for a carcinoma of the gallbladder.

A borderline indication is the presence of asymptomatic biliary sludge. Nevertheless, this condition could cause biliary pancreatitis, and in that case a cholecystectomy is usually performed in adjunct to an endoscopic retrograde sphincterotomy of the Oddi's papilla.

A prophylactic cholecystectomy is also frequently performed during total or subtotal gastrectomy and in terminal ileostomy. Both conditions, in fact, predispose the gallbladder to develop stones, in the first case because of parasympathetic denervation and in the second case because of impaired biliary metabolism.

D. F. Altomare (✉) · M. T. Rotelli
Department of Emergency and Organ Transplantation, University of Bari, Bari, Italy
e-mail: donatofrancesco.altomare@uniba.it

© Springer Nature Switzerland AG 2019
D. F. Altomare, M. T. Rotelli (eds.), *Nutritional Support after Gastrointestinal Surgery*, https://doi.org/10.1007/978-3-030-16554-3_4

4.1.2 Technique

Today almost all the cholecystectomies are performed by laparoscopic approach by three or four ports introduced into the abdomen. A single-port approach is also a feasible surgical option with the aim of further minimizing the surgical trauma and abdominal scars. The technique is well described in any textbook of surgery. Briefly, after the induction of the pneumoperitoneum with the patient in the anti-Trendelenburg position, the gallbladder is gently grasped with atraumatic forceps and attracted away from the liver toward the anterior abdominal wall in order to expose the Calot triangle to allow a safe dissection of the cystic duct and cystic artery. Once these elements have been identified they are clipped with clips and safely interrupted by scissors or other devices. The gallbladder is then attracted away from the gallbladder bed showing the areolar tissue which can be easily dissected with a hook diathermy or other devices (i.e., Ultracision forceps, Ethicon Johnson & Johnson). After the control of the hemostasis and remnant part of the cystic duct and cystic artery, the removed gallbladder is placed into a bag and extracted through the main trocar access. Despite its standardization, this operation is not always easy to perform because of the frequent anatomic variation of the biliary tree and vascular supply and because of the visceral adhesions that follow the inflammatory process of cholecystitis. Difficult cases can be removed by anterior approach starting from the fundus of the gallbladder.

4.2 Pathophysiology of GI Function After Cholecystectomy

The role of the gallbladder is to store bile and to reverse its content in the duodenum by its contraction and emptying following food and neurohormonal signals (mainly cholecystokinin) [2]. Bile is composed of bile salts, cholesterol, and bilirubin, and its role not simply favors the digestion and absorption of fatty foods and fat-soluble vitamins into the gastrointestinal tract but also regulates the equilibrium of the gut microbiota [3].

Cholecystectomy profoundly alters this fine mechanism allowing a continuous and unregulated passage of bile acids into the duodenum. A first consequence is a modified gastrointestinal motility with more frequent onset migrating motor complex causing diarrhea. Furthermore, important changes can occur in the composition of the gut microbiota [4].

Postcholecystectomy syndrome is a well-known sequela of cholecystectomy [5] with diarrhea, bloating, and abdominal pain. Finally, some studies have suggested association with the occurrence of a metabolic syndrome and liver steatosis as a result of increase enterohepatic recirculation of the bile acids.

4.3   Nutritional Support After Cholecystectomy

Genetic factors and chronic overnutrition due to a high carbohydrate and low-fiber diet both contribute to the onset of gallstone disease. This is supported by epidemiological data showing a significant association between gallbladder stones and high cholesterol level or diabetes in overweight female patients [6, 7].

After cholecystectomy, the unregulated secretion of hydrophobic secondary bile acids directly into the small bowel could impact on the gastrointestinal motility and the gut microbiota composition, resulting in nausea, bloating, diarrhea, and abdominal pain. Furthermore, epidemiological reports suggest that cholecystectomy, independently of cholelithiasis, may induce abdominal obesity, dyslipidemia, hypertension, impaired glucose tolerance, resembling a metabolic syndrome [8], elevated liver enzymes, and even cirrhosis, suggesting a profound impact of cholecystectomy on the metabolic homeostasis [9]. Based on these evidences, the dietary approach should aim to prevent the "postcholecystectomy metabolic syndrome" through both short- and long-term targeted nutritional interventions.

4.3.1   Short-Term Diet

Frequent, small, hypocaloric, and low-fat daily meals are recommended in order to ensure adequate bile acid mixture following cholecystectomy. Further, patients should be educated to slow food intake to facilitate the digestive process. Indeed, a slow oral processing provides an easier passage of food through the gastrointestinal tract and protects the gut epithelium [10] by mixing the alimentary bolus with salivary amylases, mucins, and glycoproteins.

Since after cholecystectomy the bile flow is continuous and less concentrated, fat intake needs to be moderate (i.e., cholesterol <200 mg/die) [11] to provide lipid digestion as shown by clinical studies reporting a greater tendency to develop diarrhea syndrome in patients not attending a low-fat diet after cholecystectomy [12]. In order to limit cholesterol and fat intake, butter and other animal fats, margarine, eggs, shellfish, shrimps, seasoned dairy products, processed meats, and fried or processed food should be avoided. However, fat should not be completely excluded from the postcholecystectomy diet. Instead, extra-virgin olive oil should be used for dressing and cooking, because of its role in promoting digestive processes and the absorption of fat-soluble vitamins (A, D, E, K). This is particularly important because deficiencies of A, D, E, and K vitamins could arise as a consequence of the postcholecystectomy diarrhea syndrome and the use of drugs such as cholestyramine [13].

As mentioned above, chronic diarrhea syndrome is one of the most feared post-cholecystectomy complications. An adequate fluid intake should therefore be suggested to replenish fluids lost with feces, but sometimes nausea makes water drinking difficult. Hence, fresh fruit/vegetable juices with no sugar added, vegetable broth, and tea (green or black) are suitable alternatives to fresh water and, furthermore, a good source of vitamins and minerals. Lemon juice and mint leaves could be added for their beneficial activities on both nausea and diarrhea [14]. Infusion with mint and mallow can be used to minimize bloating symptom and improve the fluid taste.

The role of fiber intake [15] has been highlighted for prevention and treatment of gallbladder disease, before and after surgery. However, when treating postcholecystectomy patients, fiber should be gradually increased starting from low amounts in order to not worse bloating and abdominal pain symptoms following surgery. Oats, rice, millet, and quinoa, vegetables such as courgettes and pumpkin, fish and small amount of nonfat meat, fruits such as banana, pineapple, blueberry, and raspberry can be proposed in the first week. Onion, garlic, mushrooms, apples, watermelon, and peaches need to be introduced later [16].

In this context, it is also important to sustain hepatic activity by introducing artichokes late in the diet. Indeed, artichokes are choleretic, cholagogue, and hepatoprotective [17]. On the other side, alcohol, preservatives, refined sugars, and chocolate should be eliminated.

Row mushrooms, pineapple, and papaya are digestive enzyme-rich food which can help to improve digestion process. In addition, such proteolytic enzymes have the ability to release active peptide sequences depending on their substrate [18]. Supplementation of oral digestive enzymes could be requested.

After cholecystectomy, the abnormal concentration of bile acids may cause alteration in gut microbiota composition through an oxidative and cytotoxic mechanism [19]. Accordingly, oral pre-/probiotics need to be supplemented in order to prevent dysbiosis and diarrhea.

4.3.2 Long-Term Diet

Although the gallbladder is not a vital organ, its removal can affect patient's health and quality of life. Several studies have documented the tendency to gain weight postcholecystectomy, mainly due to increased carbohydrate and lipid consumption, accompanied by decreased protein intake. This sort of "postcholecystectomy gluttony" has been commonly ascribed to the overall digestive and metabolic disturbances induced by the gallbladder removal. Nevertheless, it may probably be the expression of the pre-existing alimentary disturbances that ultimately caused the gallstone formation. This is a further reason for these patients to seek a long-term dietary consultation after surgery to minimize discomfort and to prevent metabolic disorders.

The postcholecystectomy long-lasting educational program should aim to:

- Control weight
- Perform physical activity
- Introduce small and frequent meals
- Introduce low fat meal
- Have high intake of soluble and insoluble fibers (whole grain, legumes, vegetables, fruits)
- Drink sulfate-bicarbonate-calcium water as a physiologic prophylactic measure to prevent further gallstone formation postcholecystectomy [20]
- Have balanced and varied diet
- Avoid refined and processed foods (white bread, white rice, packaged high-calorie snack foods)

References

1. Warttig S, Ward S, Rogers G, Guideline Development Group. Diagnosis and management of gallstone disease: summary of NICE guidance. BMJ. 2014;349:g6241. https://doi.org/10.1136/bmj.g6241.
2. Rohde U, Sonne DP, Christensen M, Hansen M, Brønden A, Toräng S, Rehfeld JF, Holst JJ, Vilsbøll T, Knop FK. Cholecystokinin-induced gallbladder emptying and metformin elicit additive glucagon-like peptide-1 responses. J Clin Endocrinol Metab. 2016;101(5):2076–83. https://doi.org/10.1210/jc.2016-1133.
3. Schubert K, Olde Damink SWM, von Bergen M, Schaap FG. Interactions between bile salts, gut microbiota, and hepatic innate immunity. Immunol Rev. 2017;279(1):23–35. https://doi.org/10.1111/imr.12579.
4. Wang W, Wang J, Li J, Yan P, Jin Y, Zhang R, Yue W, Guo Q, Geng J. Cholecystectomy damages aging-associated intestinal microbiota construction. Front Microbiol. 2018;25(9):1402. https://doi.org/10.3389/fmicb.2018.01402. eCollection 2018.
5. Moore T, Amin M. Post-cholecystectomy syndrome. Clin Pract Cases Emerg Med. 2017;1(4):446–7. https://doi.org/10.5811/cpcem.2017.6.33321. eCollection 2017.
6. Yazdankhah Kenary A, Yaghoobi Notash A Jr, Nazari M, Yaghoobi Notash A, Borjian A, Afshin N, Khashayar P, Ahmadi Amoli H, Morteza A. Measuring the rate of weight gain and the influential role of diet in patients undergoing elective laparoscopic cholecystectomy: a 6-month follow-up study. Int J Food Sci Nutr. 2012;63:645–8.
7. Ali RB, Cahill RA, Watson RG. Weight gain after laparoscopic cholecystectomy. Ir J Med Sci. 2004;173:9–12.
8. Shen C, Wu X, Xu C, Yu C, Chen P, Li Y. Association of cholecystectomy with metabolic syndrome in a Chinese population. PLoS One. 2014;9(2):e88189. https://doi.org/10.1371/journal.pone.0088189. eCollection 2014
9. Housset C, Chrétien Y, Debray D, Chignard N. Functions of the gallbladder. Compr Physiol. 2016;6(3):1549–77. https://doi.org/10.1002/cphy.c150050.
10. Boland M. Human digestion—a processing perspective. J Sci Food Agric. 2016;96(7):2275–83. https://doi.org/10.1002/jsfa.7601.
11. Riccardi; Pacione; Giacco; Rivellese. Colecistectomia in: Manuale di nutrizione clinica; Sorbona, Ed., 2009, pp. 317–18.

12. Yueh TP, Che FY, Lin TE, Chuang MT. Diarrhea after laparoscopic cholecystectomy: associated factors and predictors. Asian J Surg. 2014;37(4):171–7.
13. Van der Heide F. Acquired causes of intestinal malabsorption. Best Pract Res Clin Gastroenterol. 2016;30(2):213–24.
14. Shah AJ, Bhulani NN, Khan SH, Ur Rehman N, Gilani AH. Calcium channel blocking activity of *Mentha longifolia* L. explains its medicinal use in diarrhoea and gut spasm. Phytother Res. 2010;24(9):1392–7. https://doi.org/10.1002/ptr.3263.
15. Di Ciaula A, Garruti G, Frühbeck G, De Angelis M, De Bari O, Q-H Wang D, Lammert F, Portincasa P. The role of diet in the pathogenesis of cholesterol gallstones. Curr Med Chem. 2017; https://doi.org/10.2174/0929867324666170530080636.
16. Zannini E, Arendt EK. Low FODMAPs and gluten-free foods for irritable bowel syndrome treatment: lights and shadows. Food Res Int. 2018;110:33–41. https://doi.org/10.1016/j.foodres.2017.04.001.
17. Rangboo V, Noroozi M, Zavoshy R, Rezadoost SA, Mohammadpoorasl A. The effect of artichoke leaf extract on alanine aminotransferase and aspartate aminotransferase in the patients with nonalcoholic steatohepatitis. Int J Hepatol. 2016;2016:4030476. https://doi.org/10.1155/2016/4030476.
18. Mazorra-Manzano MA, Ramírez-Suarez JC, Yada RY. Plant proteases for bioactive peptides release: a review. Crit Rev Food Sci Nutr. 2018;58(13):2147–63. https://doi.org/10.1080/1040 8398.2017.1308312.
19. Sagar NM, Cree IA, Covington JA, Arasaradnam RP. The interplay of the gut microbiome, bile acids, and volatile organic compounds. Gastroenterol Res Pract. 2015;2015:398585.
20. Corradini SG, Ferri F, Mordenti M, Iuliano L, Siciliano M, Burza MA, Sordi B, Caciotti B, Pacini M, Poli E, Santis AD, Roda A, Colliva C, Simoni P, Attili AF. Beneficial effect of sulphate-bicarbonate-calcium water on gallstone risk and weight control. World J Gastroenterol. 2012;18(9):930–7. https://doi.org/10.3748/wjg.v18.i9.930.

Nutritional Support After Surgery of the Pancreas

5

Emanuele Felli and Sebastio Perrino

5.1 Introduction

Pancreatic cancer is an aggressive disease with a very poor prognosis, with a close parallel between disease incidence and mortality [1]. Even after potential curative resection, most patients will eventually have recurrence, and 5-year survival of completely resected patients is only up to 25% [2]. Aspecific symptoms frequently appear late, when the cancer become unresectable and can be treated only with palliative treatments to overcome the duodena obstruction of the bile flow.

No standardized screening exists, even for patients that have a family history. Surgical resection is the only potentially curative treatment, with neoadjuvant and adjuvant chemotherapy used in combination with surgery. Unfortunately, only 20% of patients are eligible to surgery at diagnosis. In Europe, pancreatic cancer is the seventh most frequent cancer with an incidence of 11.6 men out of 100,000 diagnosed with pancreatic cancer each year; differences exist when considering different countries. Pancreatic cancer is responsible of approximately 35,000 death every year. In women the incidence is lower, about 8.1 out of 100,000 women. Most of the cases are diagnosed above the age of 65, and frequency of new diagnosed cases increases with age. It is the fifth leading cause of cancer-related deaths. Pancreatic ductal adenocarcinoma is by far the most common pancreatic neoplasm.

E. Felli (✉)
Department of General, Digestive and Endocrine Surgery, University Hospital of Strasbourg, Strasbourg, France
e-mail: emanuele.felli@chru-strasbourg.fr

S. Perrino
Endocrinology Unit, Department of Emergency and Organ Transplantation, University Aldo Moro of Bari, Bari, Italy

© Springer Nature Switzerland AG 2019
D. F. Altomare, M. T. Rotelli (eds.), *Nutritional Support after Gastrointestinal Surgery*, https://doi.org/10.1007/978-3-030-16554-3_5

43

5.2 Diagnosis

No ideal screening method for pancreatic cancer exists. Early stage or premalignant lesions of pancreatic cancer usually have no clinical signs, so early diagnosis is difficult. In high-risk patients, regular follow-up can be advised. Clinical symptoms are jaundice, weight loss, and abdominal or back pain. Sometimes diabetes or pancreatitis may be the first clinical manifestation of the disease. When pancreatic cancer is suspected, abdominal ultrasound is usually the first and less expensive radiological exam. Contrast-enhanced multi-detector computed tomography (MD-CT), magnetic resonance imaging (MRI), and endoscopic ultrasound (EUS), together with magnetic resonance cholangiopancreatography (MRCP), have the highest sensitivity not only for detection of pancreatic cancer but to provide additional information on the pancreatic and the bile ducts. Endoscopic retrograde cholangiopancreatography (ECRP) is rarely used as a diagnostic method. It is usually indicated to put a biliary stent and/or associated with EUS to make core biopsy of the lesions. Tumor markers usually screened are CA 19.9 and CEA. CA19.9 can also be elevated in other diseases, and so they are not specific for pancreatic cancer. Up to 9% of the population is Lewis-negative phenotype, a blood group variation which is characterized by a non-detectable value of CA 19.9.

5.3 Treatment

Surgical resection (Fig. 5.1) is the only treatment for cure and can result in significantly longer survival compared with other treatment options. Pancreatic cancer without distant metastasis can be divided into three categories, resectable, borderline resectable, and locally advanced, according to the extent of local extension. Definitions of these categories are not uniform according to the different consensus conferences or scientific societies. For example, one of the most commonly used definitions of borderline resectable pancreatic cancer includes no distant metastases, venous involvement of the superior mesenteric vein or portal vein, gastroduodenal artery encasement up to the hepatic artery, and tumor abutment of the superior mesenteric artery of less than or equal to 180°. In specialized centers, en bloc resection of the portal vein or superior mesenteric vein is practiced with good results. However, when there is tumor abutment of the major artery such as the superior mesenteric artery, surgical resection often results in positive surgical margin. The role of neoadjuvant treatment has been investigated in many clinical trials. The rationale for neoadjuvant therapy is the better selection of patients who ideally would not benefit of surgical resection because of an aggressive disease with early recurrence or distant metastases. The possibility of completing a full dose chemotherapy regimen is also another argument in favor of a preoperative treatment. In patients with borderline resectable pancreatic cancer after effective neoadjuvant therapies, the R0 resection rate is higher, and survival of patients who underwent surgical resection is better than that of those who did not. By the way, there is no

Fig. 5.1 Schematic representation of a tumor of the head of the pancreas. Areas enclosed in the rectangles must be removed

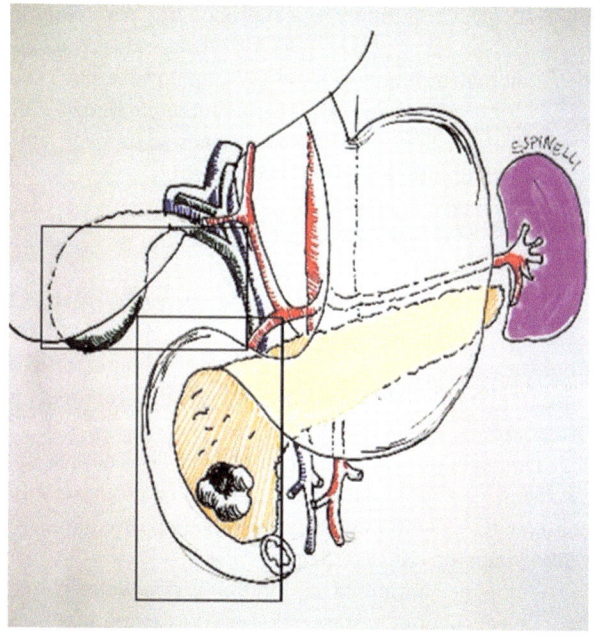

evidence from randomized controlled trials or meta-analyses to recommend neoadjuvant therapies in patients with borderline resectable or locally advanced pancreatic cancer. The role of neoadjuvant therapies in patients with resectable pancreatic cancer is another unanswered question—most of the clinical trials in the past have failed to recruit the necessary number of patients, probably because of fear of loss of the opportunity for surgical resection. Surgical resection technique includes pancreaticoduodenectomy, distal pancreatectomy with splenectomy, and total pancreatectomy. There is no evidence to support the survival advantage of extended resection including wide resections of the para-aortic lymph nodes and nerve plexus. Such extended resection is associated with compromised quality of life because of intractable diarrhea and has therefore been almost abandoned. Mortality, complications, length of hospital stay, margin status, survival, and overall cost after pancreaticoduodenectomy have been reported to be related to hospital volume. Therefore, it is recommended that pancreaticoduodenectomy should be done in specialized centers that perform a large number (>15–20) of pancreatic resections annually. Minimally invasive surgery has gained popularity in recent years, even if in pancreatic surgery, especially for pancreaticoduodenectomy, the number of laparoscopic or robotic resection has slowly increased over the years. Retrospective cohort studies have shown that laparoscopic distal pancreatectomy for cancer is not inferior to open surgery in terms of survival and can benefit patients with an earlier return to diet and a shorter hospital stay. Laparoscopic or robotic pancreaticoduodenectomy requires highly trained surgical skills, and general applications of this technique to patients with pancreatic cancer remain unwarranted.

The most common surgical procedures performed to treat pancreatic cancer are:

- Pancreaticoduodenectomy (Whipple procedure)
- Pylorus-preserving pancreaticoduodenectomy (Traverso-Longmire procedure)
- Distal splenopancreatectomy (when associated with celiac trunk resection also-called modified Appleby procedure)
- Total pancreatectomy

5.3.1   Pancreaticoduodenectomy (Whipple Procedure) (Fig. 5.2)

For tumors localized in the head, in the uncus, or in the right part of the isthmus, pancreaticoduodenectomy associated with regional lymphadenectomy is the gold standard.

This operation can be conducted by laparotomy, by laparoscopy, or with robotic assistance. Abdominal incision during laparotomy may be a bilateral subcostal incision, with or without median vertical prolongation (Mercedes incision). Xyphopubic incision can also be performed.

Complete abdominal exploration is conducted to exclude peritoneal carcinomatosis, unseen liver metastases, and gross inspection of the tumor mass. Gastrocolic ligament section is then performed. Kocher's maneuvers are then realized to mobilize the entire pancreatic head, to grossly appreciate the tumor burden and contact with the surrounding structures. It is usually done until the appearance of the left renal vein. In

Fig. 5.2 Whipple reconstruction of the gastrointestinal and biliary continuity after duodenopancreatectomy and cholecystectomy

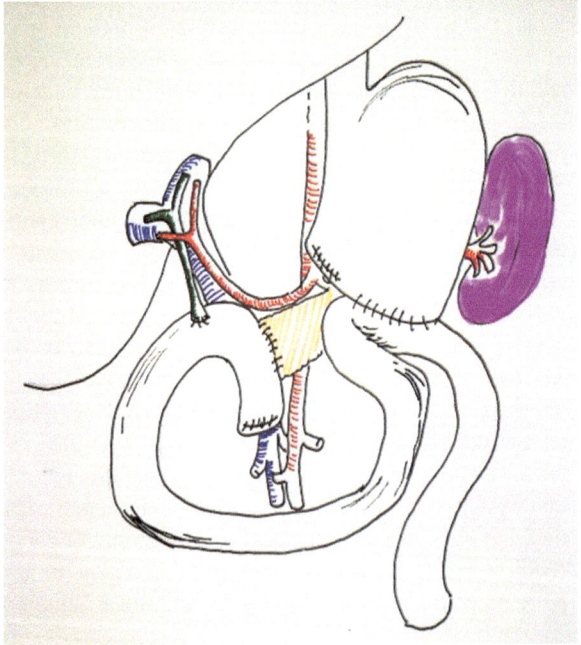

this phase picking of the inter-aortico-caval nodes for fresh frozen section is advisable. The origin of the superior mesenteric artery can be isolated at this point. Section of the Fredet fascia (inframesocolic prepancreatic-duodenal fascia) is then performed; superior mesenteric vein is then exposed to reach the pancreatic isthmus. In case of obese patients and fatty transverse mesocolon, the middle colic vein can be followed proximally toward the pancreas to find the superior mesenteric vein. The retropancreatic passage is then performed with blunt instrument. Once no contraindications to resection are found, the dissection can continue as follows: Cholecystectomy is performed (or simple cystic artery ligation and section with gallbladder dissection from his bed). Hepatic pedicle lymphadenectomy and ligation and section of pyloric artery are then done. Gastroduodenal artery is isolated and clamped with a bulldog to verify a good arterial flow from the celiac trunk (normally an arcuate ligament is diagnosed preoperatively). The artery is then ligated and cut. Celiac trunk and proximal splenic artery lymphadenectomy is then conducted. No splanchnic resection has to be made. Inferior pancreaticoduodenal vein and right gastroepiploic vein are sectioned in the Henle trunk. Two separate prolene stitches are put on the superior and inferior margin of the pancreas. The pancreas is then sectioned in the isthmus. A fresh frozen section on the pancreatic cut surface is done to find microscopic tumoral invasion. The operation continues in the inframesocolic compartment. The first jejunal loop is sectioned with a linear stapler. The Treitz ligament is sectioned, the fourth duodenal part freed, and the jejunal loop passed on the right side (*decroisement*). The retro-portal lamina is dissected along the right side of the superior mesenteric artery until its aortic origin. The common bile duct is then cut proximally. A fresh frozen section is made on the biliary stump as well. Reconstruction can be done according to Child (pancreaticojejunostomy) that can be termino-lateral, termino-terminal (with or without intussusception). The anastomosis can be done on the entire pancreatic surface or only with a Wirsung-mucosal fashion. Another common type of reconstruction is the pancreaticogastric anastomosis, with separated stitches or with a double purse string. The Wirsung duct can be tutored with an external Escat drain. Rarely a simple neoprene glue can be put in the Wirsung duct in very atrophic pancreas. Thirty/forty centimeter downward the hepatico-jejunal anastomosis is done, either with interrupted stitches or continuous running suture. In case of pancreaticogastrostomy, the bilio-enteric anastomosis is performed 4–5 cm from the jejunal stump. Finally, 30–40 cm downward the pancreatico-jejunal anastomosis is done, either manually or mechanically, usually termino-lateral. This anastomosis can be done in a pre-colic or transcolic fashion. A naso-jejunal feeding probe is then inserted and pushed distally to the gastric anastomosis. A nasogastric tube is placed in a trans-anastomotic position. One or two drainage tubes are placed around the pancreatic anastomosis and hepatico-jejunal anastomosis. The abdominal wall is the closed (Fig. 5.3).

5.3.2 Pylorus-Preserving Pancreaticoduodenectomy

The operation differs from the previous one because of the conservation of the pylorus and pyloric artery. The gastro-enteric anastomosis is usually done in an end-lateral fashion.

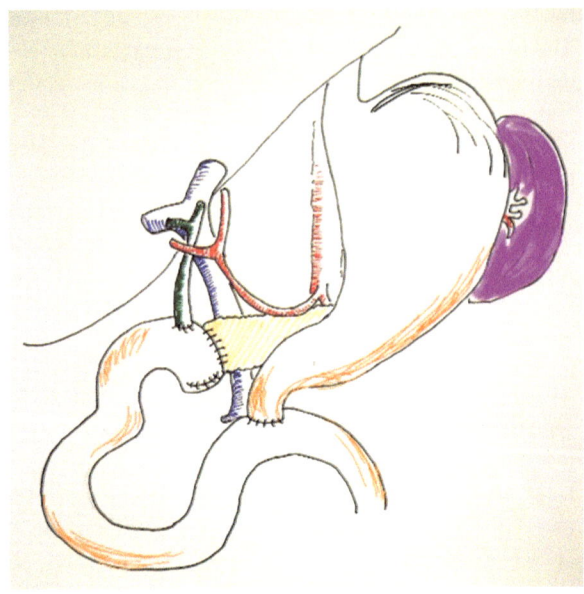

5.3.3   Distal Splenopancreatectomy

The incisions and exploration are the same. The gastrocolic ligament is opened until
the spleen with section of the left gastroepiploic artery. The superior mesenteric
vein is freed in the retro-pancreatic zone, and the splenic artery is isolated and cut
in its origin. Celiac trunk lymphadenectomy is performed. The pancreas is then cut,
and a fresh frozen section is performed on the stump. Section can be done with
stapler, energy device, or multiple prolene ligation and parenchymal section with
elective Wirsung duct section. Splenic vein is ligated and cut in the splenomesen-
teric confluence. Then the so-called RAMPS (radical antegrade modular pancreato-
splenectomy) is performed, passing on the left side of the superior mesenteric artery
vertically until the renal vein. The pancreatic body is the mobilized "en bloc" with
the surrounding tissue toward the left with the section of the origin of the transverse
mesocolon. The left colic flexure is freed and separated from the pancreatic tail. The
short gastric vessels are ligated and cut electively. The spleen is then freed from its
peritoneal connection with the diaphragm and surrounding tissues. An accurate
hemostasis is done. Gallbladder may be left in place if without stones. A drainage is
left on the pancreatic stump.

5.3.4   Modified Appleby Procedure

The original procedure was described for locally advanced gastric tumors and was
so associated with total gastrectomy. The modified procedure does not consider
total gastrectomy. The prucedure is the same as distal splenopancreatectomy but the

involvement of celiac axis need section of the common hepatic artery before gastro-duodenal artery (that is conserved). The celiac axis is then resected on its origin in the aorta. Reconstruction of celiac trunk and common hepatic artery by an inter-posed graft (saphenous vein or a prosthesis) is made by some surgeons.

5.3.5 Total Pancreatectomy

This is the same as pancreaticoduodenectomy and distal splenopancreatectomy but without section of the pancreas in the isthmus. The pancreas is resected "en bloc" with the spleen and splenic vessels. Totalization of pancreatectomy may be necessary if intraoperatively frozen sections of the pancreatic stump are positive for cancer.

5.4 Exocrine and Endocrine Consequences of Pancreatic Surgery

The pancreas has endocrine and exocrine functions. Partial or total resection of this organ may lead to a different degree of exocrine and endocrine insufficiency.

Regarding exocrine function the diminished quantity of pancreatic secretions is responsible of diarrhea (with or without steatorrhea) that can be prevented or treated with exocrine enzymes (pancreatic enzymes replacement therapy). Dietary advices and antidiarrheal drugs are sometime needed to improve the quality of life.

In the absence of duodenum, the partial gastrectomy and the proximal jejunal resection are usually responsible of a malabsorption and accelerated intestinal transit. Patients can have diarrhea as main symptom as well as abdominal pain usually improved by bowel transit. The gastroenteric anastomosis is usually asso-ciated with biliary reflux that can lead to gastritis, esophagitis, and impairment of quality of life. Typically, patients have a constant weight loss that is around 10% of the preoperative one. The development of hepatic steatosis is also usually observed in long term.

Endocrine function insufficiency is typically characterized by diabetes. Dietary advices and antidiabetic oral drugs or insulin are usually necessary. After total pan-createctomy, diabetes is constant and sometimes difficult to treat. The amount of remnant pancreas in pancreaticoduodenectomy is linked to the incidence of postop-erative diabetes.

5.5 Adjuvant Chemotherapy

Adjuvant chemotherapy for resected pancreatic cancer has been demonstrated to have better overall and disease-free survival. In general practice, adjuvant chemo-therapy with gemcitabine is given after surgery for resectable pancreatic cancer when the patient can tolerate chemotherapy. On the basis of current evidence, both FOLFIRINOX and gemcitabine plus nab-paclitaxel are the treatments of choice for patients who can tolerate these regimens. Gemcitabine monotherapy may be indicated in patients with compromised performance status. Second-line chemotherapy for metastatic pancreatic cancer following progression during first-line chemotherapy might benefit patients with good performance status and should be considered, although there is no established evidence regarding the regimen of second-line chemotherapy. The potential merits of chemoradiation therapy have been intensively studied in patients with locally advanced pancreatic cancer. However, there have been long-lasting debates about the survival benefits of chemoradiation therapy. Further investigations are needed to validate the potential survival advantages of chemoradiation therapy. Palliation may require both chemoradiation and invasive treatments such as endoscopic, percutaneous, or surgical procedures to treat life-threatening conditions like duodenal obstruction, jaundice, etc. Different strategies may be considered according to the particular clinical manifestations.

5.6   Optimal Test to Evaluate Pancreatic Exocrine Insufficiency

The most frequent clinical sign of pancreatic exocrine insufficiency (PEI) after surgery is steatorrhea, defined as presence of more than 7 g/day of fat in the stool, leading clinically to weight loss (WL) and associated generally with flatulence, bloating, urgency to stool, and cramping abdominal pain. WL may be secondary also to the onset of postsurgical diabetes, particularly after extended resections of the pancreas or in patients with underlying chronic pancreatitis (CP). Steatorrhea generally appears when greater than 90% of the typical secretion of pancreatic enzymes is lost. After PD, the combination of loss of pancreatic tissue and asynchronous mixing of pancreatobiliary secretions with the meal can lead to the onset of steatorrhea also in the presence of a more limited decrease in pancreatic enzyme secretion. Diagnosis of PEI can be difficult in practice. Pancreatic function and secretion are not solely reliant on the quantity or quality of pancreatic tissue but also depend on complex pancreatic stimulatory mechanisms. The 72-h fecal fat collection with a standard intake of fat allows the calculation of the coefficient of fat absorption; this is the gold standard test to diagnose fat malabsorption. But because this test is not available routinely, the fecal elastase-1 may be the only pancreatic function test available in clinical practice. It must be remembered that the sensitivity and the specificity of fecal elastase-1 for steatorrhea in patients after pancreatectomy is low. Steatorrhea may be present in operated patients even if the fecal elastase-1 is only mildly decreased, suggesting that steatorrhea is only partially related to an absolute decrease in digestive enzyme production and may be secondary to other mechanisms reported above. None of the above tests measure protein and carbohydrate malabsorption or discriminate between extrapancreatic factors.

5.7 Pancreatic Enzyme Replacement Therapy

Awareness of PEI by many physicians is poor outside of high-volume HPB centers and especially among physicians in primary care; consequentially, patients who present with symptoms of PEI may be overlooked or advised to adopt inappropriate dietary restrictions in an attempt to control the symptoms. A study of pancreatic cancer patients and bereaved caregivers identified that their primary unmet need was the difficulty in managing gastrointestinal problems, diet, and digestion; indeed, many of these patients and caregivers cited delays in dietary assessment and initiation of PERT as causing additional distress that could have been prevented. PERT has been shown to stabilize weight, improve dietary intake, and decrease daily stool frequency in patients with inoperable pancreatic cancer. PERT use appeared to improve survival in patients postresection in a post hoc subgroup analysis, predominantly in those with pancreatic ductal dilation. A prospective, placebo-controlled study evaluating the impact of PERT on WL in patients with unresectable pancreatic cancer did not show any benefit; however, these patients were prescribed a fixed dose of PERT with no alteration for individualized portion size or food type. A prospective study evaluating patients with unresectable pancreatic cancer undergoing chemotherapy found less decrease in BMI in patients who received PERT. Management of PEI involves replacing the lack of adequate pancreatic enzymes, which should be used to maintain weight and improve the symptoms of maldigestion. Therapy should start with doses of 40,000–50,000 units of lipase with meals and 10,000–25,000 units with every snack. Dose escalation and inhibition of gastric acid secretion may be warranted according to response; in patients who fail to respond to treatment, extrapancreatic causes should be evaluated. Dietary intake and nutritional status should be monitored regularly to maximize patient compliance and specialist dietetic assessment sought in patients with underlying malnutrition. Although one study evaluating the impact of the scheduling of PERT administration on fat malabsorption suggested the optimal timing of administration was during or after meals, no significant difference was observed when patients took the PERT immediately before meals.

5.8 Nutritional Support After Pancreatic Surgery

Each patient submitted to pancreatic surgery has individualized nutritional needs. Then, the involvement of a dietitian and/or endocrinologist to oversee dietary management is recommended. The role of the dietitian is to assess the nutritional adequacy of the patient's diet. Dietary advice can then be specifically tailored to improve energy and protein intake and to ensure that the diet is nutritionally adequate in micronutrients. The common problems that significantly impact on nutrition in patients undergoing PD include:

(a) Weight loss
(b) Poor appetite and feeling full quickly

(c) Changes in taste and smell

(d) Diarrhea or other changes in bowel habits

(e) Poor digestion (maldigestion) and absorption (malabsorption) of fats and proteins caused by lack of pancreatic enzymes

This chapter provides general suggestions about how to manage these problems. Thus, what the patients can eat and drink will depend on own individual circumstances.

5.8.1 Weight Loss: Nutritional Supplements

Weight loss in adults or lack of weight gain in children is common in pancreatic exocrine insufficiency (PEI) because of fat malabsorption and the patient's fear of eating (due to exacerbation of symptoms such as abdominal pain and flatulence). Ensuring adequate growth in children and preventing weight loss in adults are paramount. Historically, dietary fat intake has been restricted in patients with PEI to minimize fat malabsorption and reduce steatorrhea. However, low-fat diets are lower in total energy content, and restricting fat intake also reduces intake of fat-soluble vitamins, which are already malabsorbed in people with PEI. Today, fat restriction is no longer recommended [3]. Normal- and high-fat diets have been successfully used in combination with adequate pancreatic enzyme replacement therapy (PERT) [3].

If the patient can't eat a balanced diet or is losing too much weight, the endocrinologist or dietitian may suggest nutritional supplements such as Ensure, Fortisip, or Resource (one or two times per day). These nutritional supplements contain carbohydrates, proteins, and other nutrients (calcium, iron, vitamins C and B). Glucose powder supplements provide little nutrition and are rarely recommended [3, 4]. These nutritional supplements should be taken in addition to the patient's usual meals, i.e., as snacks between meals. The above nutritional supplements are available in ready-made as drinks or bars or in powdered form that can be mixed with milk, water, and food.

5.8.2 Poor Appetite and Feeling Full Quickly

Food intake may be better distributed across six or more smaller meals throughout the day rather than three large meals. Large meals may not be appetizing to a patient with symptomatic PEI, and small meals are often better tolerated. It's important to eat small meals frequently, e.g., every 2–3 h, and ensure that meals and snacks are nourishing and include protein, e.g., meat, chicken, fish, dairy products, eggs and nuts. The patients have to choose nourishing drinks such as milk. Nutritional supplement drinks may be prescribed after surgery. The dietitian should recommend to add milk powder to cereals, sauces, desserts, mashed vegetables, soup, drinks, and egg dishes. Further, it is useful to add cheese to sauces, soup, baked beans, vegetables, casseroles, salads, and egg dishes and add golden syrup or honey to cereal, fruit, and drinks. The mixing of chyme with pancreatic enzymes is considered more efficient when smaller meals are consumed. This regimen may improve the energy, protein, and micronutrient content of the diet and therefore facilitate weight gain and nutritional improvements. With any PEI, alcohol abstinence is crucial, as alcohol inhibits gastric lipase secretion and therefore contributes to fat malabsorption. Indeed, with time, alcohol consumption can cause more severe and rapid deterioration of pancreatic function.

5.8.3 Changes in Taste and Smell

Research suggests many cancer patients experience changes to their taste and smell, causing anxiety, nutritional complications, and quality-of-life indications. According to a review published in 2009 in *The Journal of Supportive Oncology*, slightly more than two-thirds (68%) of cancer patients undergoing chemotherapy reported a change in sensory perception, including a reduction or loss in taste sensitivities or a metallic taste in the mouth [5]. Another 2010 study published in the peer-reviewed journal *The Oncologist* found that nearly 70% of patients receiving chemotherapy reported an altered sense of taste [6]. Some people find that they like to enhance the taste of food using spices, herbs, lemon, lime, ginger, garlic, honey, chilli, pepper, vinegar, soy sauce, or pickles, while others find avoiding strong flavors and eating bland, unseasoned food helpful. The dietitian should provide some tips for coping with changes in taste and smell. For instance, to marinate meat or vegetables to add more flavor, or if the food tastes too sweet, add lemon juice or salt—starting with a few drops, increasing until you find the taste acceptable. If food tastes metallic or too salty, add sugar or honey. Some people find a drinking straw helpful to bypass your taste buds. Another useful recommendation is to choose foods without a strong smell—sometimes the smell of a food can put you off eating. Cold foods tend not to smell as much as hot foods. Sometimes it can be helpful to eat with plastic utensils to reduce the metallic taste and avoid drinking out of aluminum cans and storing food in metallic containers.

5.8.4 Diarrhea or Other Changes in Bowel Habits

Diarrhea is an unpleasant but common side effect in people receiving treatment for pancreatic cancer. However, diarrhea may also be caused by the cancer itself, and it can sometimes be a sign of something more serious [3, 4]. If the diarrhea doesn't seem severe, but starts to interfere with daily activities of the patient, such as if the subject is concerned about leaving home or going somewhere without a toilet nearby, it is possible take action by modifying what he is eating and drinking. The dietitian should recommend to switch to a liquid-based diet, such as apple juice, and clear broth and avoid milk products. Further, when the diarrhea begins, the patient has to eat low-fiber foods and eat five to six small meals a day. Avoid foods that can irritate the digestive tract such as dairy products, spicy foods, alcohol, and high-fat foods and beverages that contain caffeine, orange juice, or prune juice. It should be recommended to eat yogurt supplementing diet with probiotics. The probiotics are beneficial bacteria that may help restore normal digestion. Lactobacillus and bifidobacterium are two examples of probiotics.

5.8.5 Poor Digestion (Maldigestion) and Absorption (Malabsorption) of Fats and Proteins Caused by Lack of Pancreatic Enzymes

The most frequent clinical sign of pancreatic exocrine insufficiency (PEI) after surgery is steatorrhea, defined as presence of more than 7 g/day of fat in the stools, leading clinically to weight loss (WL) and associated generally with flatulence, bloating, urgency to stool, and cramping abdominal pain [7]. WL may be secondary also to the onset of postsurgical diabetes, particularly after extended resections of the pancreas or in patients with underlying chronic pancreatitis (CP) [8]. Steatorrhea generally appears when greater than 90% of the typical secretion of pancreatic enzymes is lost [7, 8]. After pancreaticoduodenectomy (PD), the combination of loss of pancreatic tissue and asynchronous mixing of pancreato-biliary secretions with the meal can lead to the onset of steatorrhea also in the presence of a more limited decrease in pancreatic enzyme secretion [8]. Diagnosis of PEI can be difficult in practice. Awareness of PEI by many physicians is poor outside of high-volume HPB centers and especially among physicians in primary care; consequentially, patients who present with symptoms of PEI may be overlooked or advised to adopt inappropriate dietary restrictions in an attempt to control the symptoms. A study of pancreatic cancer patients and bereaved caregivers identified that their primary unmet need was the difficulty in managing gastrointestinal problems, diet, and digestion; indeed, many of these patients and caregivers cited delays in dietary assessment and initiation of pancreatic enzyme replacement therapy (PERT) as causing additional distress that could have been prevented [8]. PERT has been shown to stabilize weight, improve dietary intake, and decrease daily stool frequency in patients with inoperable pancreatic cancer. PERT use appeared to improve survival in patients postresection in a post hoc subgroup analysis, predominantly in those with pancreatic ductal dilation. A prospective, placebo-controlled study evaluating the impact of PERT on WL in patients with unresectable pancreatic cancer did not show any benefit; however, these patients were prescribed a fixed dose of PERT with no alteration for individualized portion size or food type. A prospective study evaluating patients with unresectable pancreatic cancer undergoing chemotherapy found less decrease in BMI in patients who received PERT [8]. Management of PEI involves replacing the lack of adequate pancreatic enzymes, which should be used to maintain weight and improve the symptoms of maldigestion. Therapy should start with doses of 40,000–50,000 units of lipase with meals and 10,000–25,000 units with every snack [8, 9]. Dose escalation and inhibition of gastric acid secretion may be warranted according to response; in patients who fail to respond to treatment, extrapancreatic causes should be evaluated. Dietary intake and nutritional status should be monitored regularly to maximize patient compliance and specialist dietetic assessment sought in patients with underlying malnutrition. Although one study evaluating the impact of the scheduling of PERT administration on fat malabsorption suggested the optimal timing of administration was during or after meals, no significant difference was observed when patients took the PERT immediately before meals.

References

1. Siegel R, Ma J, Zou Z, et al. Cancer statistics, 2014. CA Cancer J Clin. 2014;64:9–29.
2. Kamisawa T, Wood LD, Itoi T, Takaori K. Pancreatic cancer. Lancet. 2016;388:73–85.
3. Gilliland TM, Villafane-Ferriol N, Shah KP, Shah RM, Tran Cao HS, Massarweh NN, et al. Nutritional and metabolic derangements in pancreatic cancer and pancreatic resection. Nutrients. 2017;9(3):E243.
4. Hong JH, Omur-Ozbek P, Stanek BT, Dietrich AM, Duncan SE, Lee YW, et al. Taste and odor abnormalities in cancer patients. J Support Oncol. 2009;7(2):58–65.
5. Zabernigg A, Gamper EM, Giesinger JM, Rumpold G, Kemmler G, Gattringer K. Taste alterations in cancer patients receiving chemotherapy: a neglected side effect? Oncologist. 2010;15(8):913–20.
6. Othman MO, Harb D, Barkin JA. Introduction and practical approach to exocrine pancreatic insufficiency for the practicing clinician. Int J Clin Pract. 2018;72(2) https://doi.org/10.1111/ijcp.13066.
7. Ramsey ML, Conwell DL, Hart PA. Complications of chronic pancreatitis. Dig Dis Sci. 2017;62(7):1745–50.
8. Singh VK, Haupt ME, Geller DE, Hall JA, Quintana Diez PM. Less common etiologies of exocrine pancreatic insufficiency. World J Gastroenterol. 2017;23(39):7059–76.
9. Struyvenberg MR, Martin CR, Freedman SD. Practical guide to exocrine pancreatic insufficiency—breaking the myths. BMC Med. 2017;15(1):29.

Nutritional Support After Surgery of the Small Bowel

6

Donato Francesco Altomare and Maria Teresa Rotelli

6.1 Indication

The most frequent diseases involving the small bowel and requiring surgery are Crohn's disease and postoperative or post inflammatory adhesion. Less frequently small bowel resection is performed because of ischemia for mesenteric artery thrombosis or embolism and for neoplastic diseases arising in the small bowel (rare) or compressing or infiltrating the small bowel from outside (desmoid tumors, sarcomas, lymphomas, etc.). Furthermore, the small bowel is often interested by penetrating wounds (stab wounds or gunshot wounds) or can be strangulated in inguinal hernia, ventral hernia, or volvulus. Another rare small bowel disease requiring surgery is intussusception often associated to pedunculated polyps.

Crohn's disease is a chronic inflammatory disease of the bowel, particularly frequent in north Europe and the USA but frequent also in other industrialized countries, which can involve any part of the GI tract.

The small bowel involvement (particularly the last ileal loop) is the most frequent one. The disease distribution is typically segmental with involvement of all the layers of the viscera, leading to strictures, abscesses, and fistulae with other organs or with the skin.

Postoperative adhesion formation is a physiologic event following surgical trauma or peritoneal inflammations. Both can cause extensive intraperitoneal exudate collection containing fibrin and other plasmatic factors responsible for collagen formation. This event is favored by the inhibition of the mesothelial cells to produce uPA (urokinase plasminogen activator) because of the trauma or inflammation. Postoperative adhesions are often responsible for small bowel occlusion or sub-occlusion requiring emergency surgery.

D. F. Altomare (✉) · M. T. Rotelli
Department of Emergency and Organ Transplantation, University of Bari, Bari, Italy
e-mail: donatofrancesco.altomare@uniba.it

© Springer Nature Switzerland AG 2019
D. F. Altomare, M. T. Rotelli (eds.), *Nutritional Support after Gastrointestinal Surgery*, https://doi.org/10.1007/978-3-030-16554-3_6

Intestinal infarction due to vascular occlusion usually involves a large part of the small bowel and is an emergency condition with very severe prognosis. In patients surviving the operation, a proximal jejunostomy is often fashioned with very high output requiring total parenteral nutrition or a small bowel transplantation.

Traumatic injuries of the small bowel usually require closure of the bowel lesions or limited resection without significant changes in the physiology of this organ.

6.2   Surgical Techniques

Resection of a small bowel segment involves ligation of the relative mesenteric arteries and veins and interruption of the bowel. Intestinal continuity is then restored by an end-to-end or end-to-lateral mechanical or hand-sew anastomosis.

Surgery for Crohn's disease deserves some more technical information. In fact, due to its segmental distribution, the frequent recurrence of the disease, and the young age of these patients, surgery for Crohn's disease must be as conservative as possible because repeated bowel resection can lead to a short bowel syndrome with severe malabsorption. A way to avoid small bowel resection is the stricturoplasty which should be performed whenever possible. There are several types of stricturo-plasties depending on the extent of the stricture and its location. The most common one is the Heineke-Mikulicz stricturoplasty (Fig. 6.1) to treat short intestinal seg-ments with stricture. It consists into a longitudinal opening of the stricture and its transverse closing.

When the length of the stricture is longer than 8–10 cm, a Finney stricturoplasty (Fig. 6.2) is fashioned. In this operation the antimesenteric side of the bowel with the stricture is incised, and the intestinal walls are resutured in order to create an ileal pouch.

In case of multiple segmental strictures or single stricture longer than 10 cm, Michelassi's stricturoplasty (Fig. 6.3) could be fashioned.

Finally, an ileocolic stricturoplasty (Fig. 6.4) can be done instead of ileocecal resection in case of stricture involving the last ileal loop.

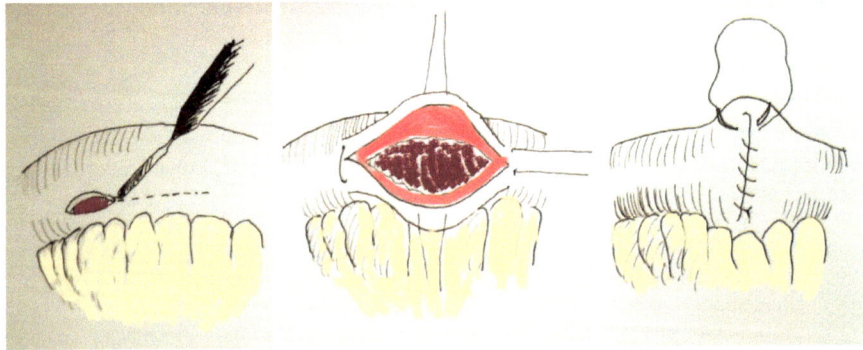

Fig. 6.1   Heineke-Mikulicz stricturoplasty

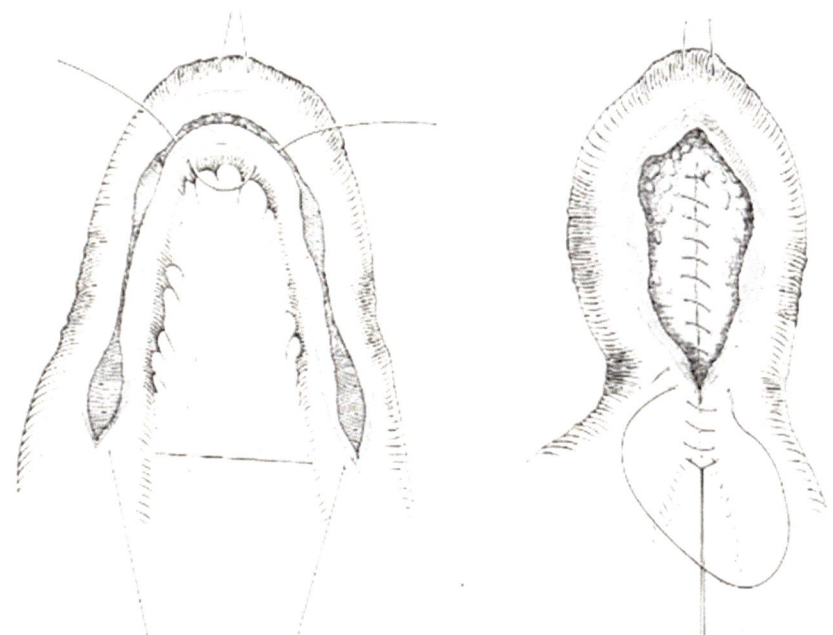

Fig. 6.2 Finney stricturoplasty

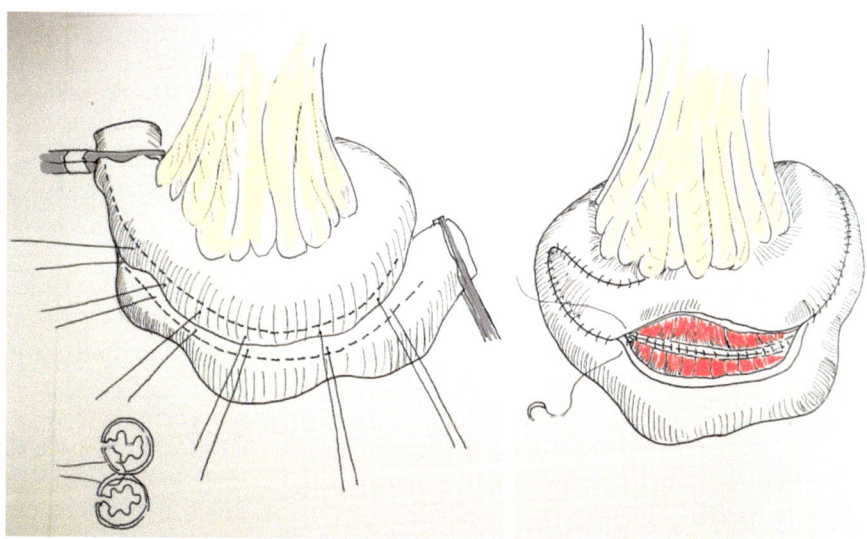

Fig. 6.3 Michelassi's stricturoplasty

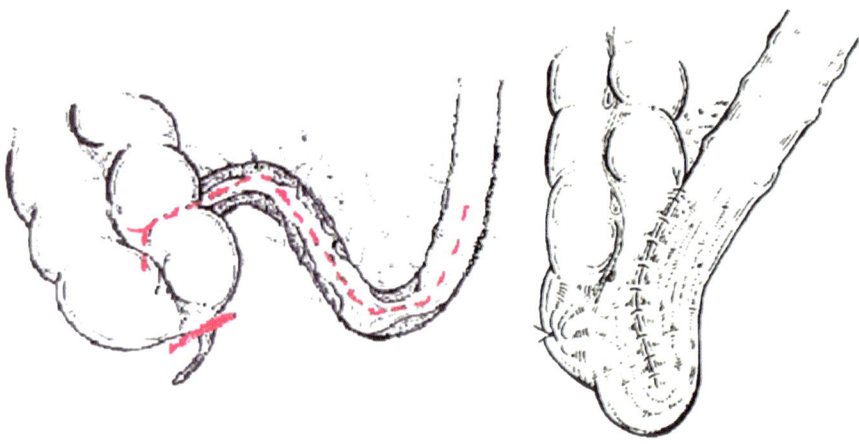

Fig. 6.4 Ileocolic stricturoplasty

6.3   Pathophysiologic Implications

Digestion and absorption of nutrients in the gut are complex processes involving the nutrients characteristics, the digestive enzymes, the bile, and the gastrointestinal mucosa. Furthermore, gut motility, regulating the transit of intraluminal content, and the secretion of digestive enzymes are regulated by complex hormonal, neural (enteric and autonomic nervous system), and local factors.

In particular the presence of nutrients themselves in the small bowel is able to stimulate the contraction of the gallbladder and the secretion of pancreatic enzymes. At the same time, the pattern of small bowel motility is converted from fast to fed type.

Segmental or limited ileal resection (<50 cm) usually does not cause significant changes in the ability to complete the digestion and to absorb nutrients by the small bowel and therefore does not require special nutritional support after the first month when the healing of the anastomosis is completed.

Nevertheless, a larger removal of the ileum can severely compromise the ileal brake mechanism which is an "intraluminal nutrient-triggered feedback control from the distal to the proximal gut." This mechanism is stimulated in particular by fat-containing foods [1, 2].

Such a nutrient-triggered inhibitory feedback arising in the more distal to the proximal gut is a well-known mechanism present in other parts of the GI tract (duodenal brake inhibits gastric emptying and colonic brake delay ileal motility) [3, 4].

Complex interaction of several gut peptides, including glucagon-like peptide 1 (GLP-1 synthesized by the endocrine L cells in the ileum and colon), neurotensin (produced by the mucosal endocrine N cells), and peptide YY (PYY, secreted by L cells in the distal ileum and colon), in combination with intrinsic and extrinsic (particularly sympatho-adrenergic nerves) neural pathways seem to be implicated in the control of the ileal brake.

In patients suffering of Crohn's disease, the post-operative nutritional support should take into consideration not only the type of surgery performed but also the malnutrition and food intolerance often present in these patients. Furthermore, most of these patients are under heavy medical treatments including cortisone, anti-inflammatory drugs, cytostatic drugs, and anti-TNFα drugs.

In the early postoperative period, a diet with low residual fiber is suggested in order to facilitate the transit through the recent anastomosis.

However, it should be remembered that when the ileocecal valve has been removed and an ileocolonic anastomosis has been fashioned, the risk of ileal bacterial overgrowth is high, contributing to dyspepsia, abdominal bloating, and diarrhea [5].

For all these reasons, if the small bowel resection larger than 50 cm, or involving the jejunum, has been performed, and the inhibitory feedback of the ileal brake is impaired or absent, gastric emptying and small intestinal transit will be accelerated, leading to an increased amount of undigested and unabsorbed nutrients in the colon. Such condition can explain the onset of symptoms such as diarrhea and malabsorption commonly present in inflammatory bowel diseases and after small intestine resection, requiring specialized nutritional advices.

Finally, patients submitted to extensive small bowel resection for intestinal infarction, strangulation, or desmoid tumor invading the mesenterium develop a severe short bowel syndrome requiring total parenteral nutrition and intestinal transplantation.

6.4   Nutritional Support After Surgery on the Small Bowel

The small bowel, whose length in adult ranges from 300 to 800 cm, is divided in duodenum (25–30 cm), jejunum (160–200 cm), and ileum. Carbohydrates, protein, and water-soluble vitamins are absorbed mostly in its proximal tract (duodenum and jejunum), whereas fat-soluble vitamins, vitamin B12 (bound to intrinsic factor), and fats (bound to bile salts) are absorbed in the distal tract (ileum). Ileum and large bowel are the main sites of fluid and electrolyte absorption [6] (Fig. 6.5).

Extensive resection, i.e., more than 50%, frequently disrupts the absorptive process of nutrients and fluids, generating the short bowel syndrome (SBS). Depending on the extent and site of resection, this syndrome encompasses different symptoms, including diarrhea, cramping, and abdominal pain, associated with inadequate digestion and/or absorption of nutrients and fluids. Malabsorption always requires short- or long-term home parenteral nutrition that will not be discussed in this section.

The "intestinal adaptation" is a postoperative adaptive process that compensates the reduction of the absorptive surface area that takes up to 1 or 2 years to develop and is favored by the patient's ability to resume the oral diet.

Dietary management of small bowel-resected patients should minimize symptoms and restore nutrient and fluid deficiencies. An individualized diet plan is

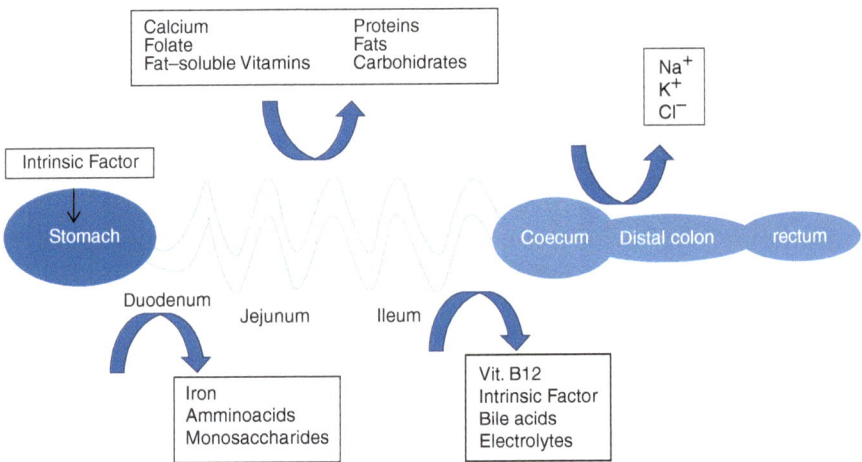

Fig. 6.5 Location of nutrients absorption into the small bowel

mandatory, to tailor the single patient food tolerance and specific time-related dietary needs, and a periodic dietary evaluation is strictly recommended. Importantly, the nutritional support should be tailored to patients' lifestyle, in order to improve the overall outcome.

Despite the important macro- and micronutrient absorption in the first 150 cm of the intestine, patients with resections of the duodenum or terminal ileum have worst prognosis than patients with jejunum loss, in terms of energy intake and nutritional deficiencies, because the duodenum and ileum have greater adaptive and motility capacities as compared with the jejunum [6, 7]. Accordingly, patients undergoing ileal resections usually need more attention in the dietary and pharmacological management. Diarrhea and steatorrhea, due to unabsorbed long chain fatty acids and bile salts, reaching colon surface, can lead to fat-soluble vitamins, vitamin B12, calcium, and magnesium deficiencies that should be eventually supplemented.

To prevent malnutrition, a high caloric diet is strongly suggested. A gradual increase of energy intake from 30 to 60 kcal/kg per day could be necessary (Khursheed), since approximately the 50% of energy introduced with diet is not absorbed in the SBS (Nightingale). If the colon is present, complex carbohydrates, fats, and proteins consumption should represent the 50–60%, the 20–30%, and the 20% of the energy intake, respectively. If colon is absent, less than half of calories should derive from complex carbohydrates and 30–40% from fats [7]. Daily caloric requirement should be distributed into six to eight small meals and prepared mainly with fresh food, whereas the intake of processed foods, instead, should be limited.

Simple sugars (sucrose and fructose, chewing gum, candies, fruit juices, sweet foods, sugar alcohols) should be avoided, because of their pro-inflammatory activity and their role in attracting water into the gut, resulting in increased diarrhea and transit time. Consumption of complex carbohydrate as rice, pasta, and bread (i.e., with spelt or khorasan wheat flour) is generally well tolerated.

A moderate consumption of food containing soluble fibers such as oatmeal, barley, peeled potatoes, fruits (strawberries, blueberries, bananas), or peeled fruits (apples, pears) help to slow down gastric emptying and the overall intestinal transit time, improving the absorption of nutrients from the unresected intestine. Moreover, soluble fibers' ability to attract water, turning into gel, also improves stool consistency. Colon fermentation of unabsorbed carbohydrates and soluble fiber into SCFA (short-chain fatty acids) could generate up to 500 and 1000 kcal per day, contributing to the daily calorie intake [8].

Leafy green vegetables such as spinach, kale, collards, turnip greens, or whole foods that are all rich in insoluble fibers should be avoided.

Strained legumes (peas, chickpeas, lentils, bean), if tolerated, can be gradually introduced in the diet [7]. Especially in patients with no colon, ingestion of 10–15 g of sodium chloride per meal could be necessary also to promote absorption of dietary carbohydrate, if no restrictions due to other pathologies are present [6] since a mechanism based on the presence of a coupled sodium/glucose molecule transport system in fact is active on intestinal brush border.

Patients should be also encouraged to consume appropriate amounts of high biologic value proteins from fish, meat (chicken, turkey, and beef), eggs (white eggs) in each meal. Lean cuts of meat (bresaola, chicken, or turkey breast) should be preferred in order to not exceed fat intake.

Although fats represent a significant energy source, its restriction could be necessary particularly in patients with colon, because of possible onset of severe steatorrhea, and oxalate nephrolithiasis. As previously reported if the colon has been removed, introduction of fat foods should not exceed 40% of total calorie intake. In both cases, in order to prevent essential fatty acid deficiency, diet should include oil with essential fatty acids (extra virgin olive oil, sesame oil, soy oil, fish). In patients with colon, intake of medium chain triglycerides (MCT) could represent an additional source of calories.

The educational program should sensitize patients to the importance of fluid oral intake and to rapidly recognize and prevent dehydration symptoms. Of note, drinking large amounts of plain water does not ensure a correct hydration status, and accordingly, fruit juices, fruit drinks, and sweet teas should be avoided. Solutions with lower sodium concentration or with high-glucose concentration without sodium may lead increased stool output, with dehydration and electrolyte stool loss. Therefore, homemade or commercially glucose-electrolyte oral rehydration solution (ORS), differing in composition, could be used. An inadequate hydration, together with an increased absorption of oxalate in Crohn's patients with the colon, could cause oxalate kidney stones formation. A low-oxalate diet, limited in wheat germ, rhubarb, beets, collards, spinach, tomatoes, and tofu, could be eventually proposed [8, 9].

Finally, a low-fermentable oligosaccharide, disaccharide, monosaccharide, and polyol (FODMAP) diet could be eventually considered if typical dietary advices fail to minimize postsurgery symptoms (diarrhea, bloating) or for patients submitted to extensive small bowel resection for Crohn's disease. FODMAPs are oligosaccharides (fructans, galacto-oligosaccharides), disaccharides (lactose), monosaccharides (fructose), and polyols (e.g., sorbitol, mannitol) that are not completely absorbed in the small intestine but are fermented in the colon, leading to an increase in intestinal water and gas production, including hydrogen and methane. FODMAPs are found in a wide range of unprocessed and processed foods and include some fruits, vegetables, dairy products and legumes, but also candies, chewing gum, and sweetener-added food.

Although the effects of a low FODMAP diet have been well investigated for irritable bowel syndrome a randomized, single-blind, crossover feeding study in ileostomy patients showed a 20% of effluent water increase after 4 days of very high FODMAP diet intake. Anyway, a low-FODMAP prescription diet needs to be supervised to prevent or manage its potentially unfavorable effects on the microbiota and its metabolites [10, 11].

References

1. Van Citters GW, Lin HC. The ileal brake: a fifteen-year progress report. Curr Gastroenterol Rep. 1999;1:404–9.
2. Maljaars PW, Peters HP, Mela DJ, Masclee AA. Ileal brake: a sensible food target for appetite control. A review. Physiol Behav. 2008;95(3):271–81.
3. Shahidullah M, Kennedy TL, Parks TG. The vagus, the duodenal brake, and gastric emptying. Gut. 1975;16(5):331–6.
4. Lin AY, Dinning PG, Milne T, Bissett IP, O'Grady G. The "rectosigmoid brake": review of an emerging neuromodulation target for colorectal functional disorders. Clin Exp Pharmacol Physiol. 2017;44(7):719–28. https://doi.org/10.1111/1440-1681.12760. Review
5. Roland BC, Ciarleglio MM, Clarke JO, Semler JR, Tomakin E, Mullin GE, Pasricha PJ. Low ileocecal valve pressure is significantly associated with small intestinal bacterial overgrowth(SIBO). Dig Dis Sci. 2014;59(6):1269–77. https://doi.org/10.1007/s10620-014-3166-7.
6. Jeejeebhoy KN. Short bowel syndrome: a nutritional and medical approach. CMAJ. 2002;166(10):1297–302.
7. Matarese LE, O'Keefe S, Kandil HM, Bond G, Costa G, Abu-Elmagd K. Short bowel syndrome: clinical guidelines for nutrition management. Nutr Clin Pract. 2005;20(5):493–502.
8. Seetharam P, Rodrigues G. Short bowel syndrome: a review of management options. Saudi J Gastroenterol. 2011;17(4):229–35. https://doi.org/10.4103/1319-3767.82573.
9. Parrish CR, DiBaise JK. Managing the adult patient with short bowel syndrome. Gastroenterol Hepatol (N Y). 2017;13(10):600–8.
10. Staudacher HM, Whelan K. The low FODMAP diet: recent advances in understanding its mechanisms and efficacy in IBS. Gut. 2017;66(8):1517–27. https://doi.org/10.1136/gutjnl-2017-313750.
11. Cox SR, Prince AC, Myers CE, Irving PM, Lindsay JO, Lomer MC, Whelan K. Fermentable carbohydrates [FODMAPs] exacerbate functional gastrointestinal symptoms in patients with inflammatory bowel disease: a randomised, double-blind, placebo-controlled, cross-over, re-challenge trial. J Crohns Colitis. 2017;11(12):1420–9. https://doi.org/10.1093/ecco-jcc/jjx073.

Nutritional Support in Patients with Intestinal Stoma

Donato Francesco Altomare and Antonio Finaldi

7.1   Indications

Intestinal stoma (from the Greek language στομα = mouth) formation (ostomy surgery) is a surgical procedure intended to divert the intestinal content outside the body through the abdomen. Due to its profound impact on the patients' emotional status and on their quality of life, a detailed preoperative counselling is mandatory, while the adhesion to strict technical rules is fundamental to prevent the frequent stoma complications.

Intestinal stomas can be divided into ileal and colonic according to the exteriorized viscera. Both can be temporary or definitive.

Temporary stomas are usually constructed to prevent the passage of the feces through a risky anastomosis or in case of peritonitis, intestinal occlusion, or penetrating trauma, when the construction of an intestinal anastomosis is discouraged by the high risk of leakage. Further indications are rectovaginal fistula, recto-vesical fistulas, and complex perianal fistulas (cryptoglandular or Crohn's related).

Definitive stomas are prepared whenever the restoration of the intestinal continuity is prevented by an advanced cancer or by the absence of the rest of the intestine (i.e., complete removal of the anus/rectum, total proctocolectomy for inflammatory disease). Even an intractable fecal incontinence might require a terminal colostomy.

D. F. Altomare (✉)
Department of Emergency and Organ Transplantation, University of Bari, Bari, Italy
e-mail: donatofrancesco.altomare@uniba.it

A. Finaldi
Centro Disturbi Alimentari Lucera Hospital (FG) ASL Foggia, Foggia, Italy

© Springer Nature Switzerland AG 2019
D. F. Altomare, M. T. Rotelli (eds.), *Nutritional Support after Gastrointestinal Surgery*, https://doi.org/10.1007/978-3-030-16554-3_7

7.2   Colostomy

Terminal (end) colostomy is carried out in cases of low-ultralow rectal or anal cancer and is usually prepared in the left iliac region using the residual sigmoid colon. In this case the stoma is usually permanent. A terminal colostomy, however, is often fashioned in case of bowel occlusion for advanced rectal or pelvic cancer or peritonitis due to diverticulitis (Hartmann's procedure) or for penetrating traumas. In these cases, the bowel continuity is normally restored by a new intraabdominal operation.

Loop (or lateral) colostomy is usually performed to protect a low rectal anastomosis, and, in this indication, it is a temporary stoma but sometimes could be fashioned as a permanent stoma in cases of inoperable pelvic cancer involving the rectum. A loop colostomy is normally positioned in the left iliac region using the sigmoid or the left colon, but in case of necessity it can be performed on the transverse colon, and the stoma is located in the epigastric region.

Coecostomy (prepared in the right iliac region) is a further modality of colonic stoma, but its use has been abandoned because of the difficulty in its management and the incomplete fecal diversion. Its formation is sometimes justified in emergency to manage an extreme dilatation of the cecum in cases of distal colon occlusion with a continent ileocecal valve (Fig. 7.1).

7.3   Ileostomy

Terminal (end) ileostomy (Brooke ileostomy) can be located on the right or left iliac abdominal quadrants. It is usually performed after a total colectomy or proctocolectomy (Fig. 7.2).

A specialized continent ileostomy is the Kock pouch ileostomy which was designed to create a reservoir and a sphincter-like mechanism to help these patients

Fig. 7.1 Prolapsing coecostomy

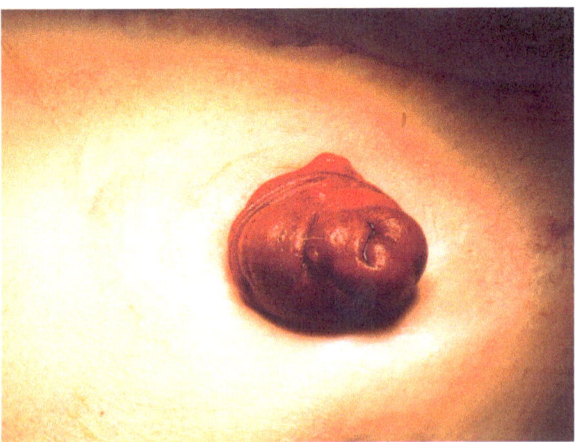

Fig. 7.2 Brooke ileostomy

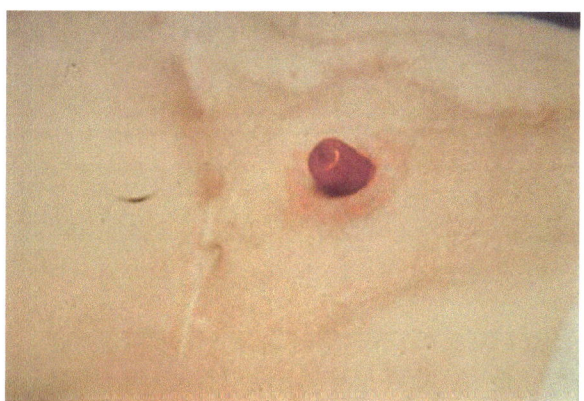

Fig. 7.3 Kock pouch continent ileostomy

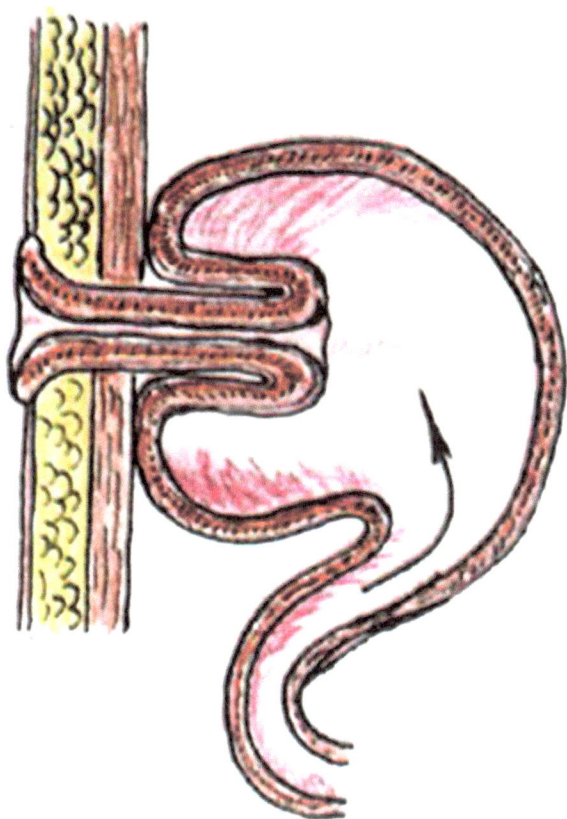

to have less fluid feces and to decrease the number of defecation through the stoma per day (Fig. 7.3).

Loop (lateral) ileostomy is usually located on the right iliac quadrant of the abdomen (Fig. 7.4).

Fig. 7.4 Mature loop
(lateral) ileostomy

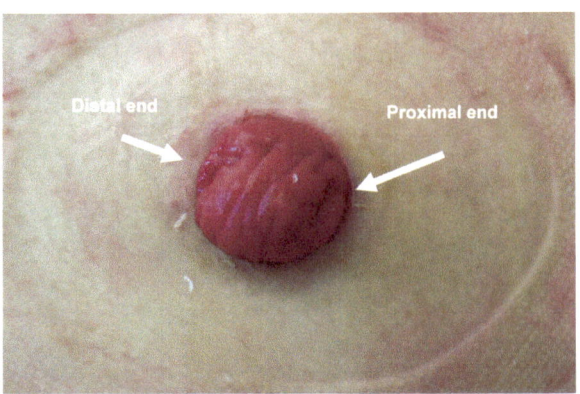

7.4   Surgical Technique

A **colostomy** can be performed by an open or a laparoscopic approach under general anesthesia. Its position on the abdomen is usually decided and marked on the abdomen before the operation taking into account the distance from bone protrusions (ribs, superior iliac spine) and scars (including the umbilicus) and the mobility of the fatty belly, if present.

During the operation the colon to be exteriorized is fully mobilized, and the vascular supply is carefully preserved. The mobilized segment of the colon is then transposed outside the abdominal wall through an adequate hole obtained by removal of a circular segment of the skin and the subcutaneous fat and by opening the fascia and stretch apart the abdominal muscles. The passage should not be too narrow to cause ischemia, but not too large to favor the onset of a parastomal hernias. In cases of loop (lateral) colostomy, the mesentery is perforated in an avascular portion close to the bowel, and the intestine is exteriorized with the help of a flexible tube passed through the mesenteric hole (Fig. 7.5).

The bowel is kept outside the abdomen by a dedicated plastic stick for a period of 5–7 days until the formation of adhesions between the bowel and the abdominal wall in order to prevent the introflexion of the stoma (Turnbull technique). In cases of permanent loop colostomy, a skin bridge can be used instead of the plastic stick (Mixter's or Quenu's colostomy) (Fig. 7.6).

The exteriorized colonic wall is sutured to the muscular fascia by at least four (on the cardinal points) resorbable sutures. The anterior wall of the colon is then opened by a diathermy with a transverse incision close to the distal part of the exteriorized colon, and the mucosa of the proximal colon is estroflexed and fixed to the derma by four to six resorbable sutures which enclose also the seromuscular layers of the colon. The procedure is repeated on the distal (dysfunctioned) part of the colon.

In case of terminal colostomy, the procedure is essentially the same. In this case the bowel is resected using a linear stapler, and while the distal part of the bowel is left closed inside the body, the proximal segment of the bowel is exteriorized and

Fig. 7.5 Early lateral ileostomy kept in place by a stick

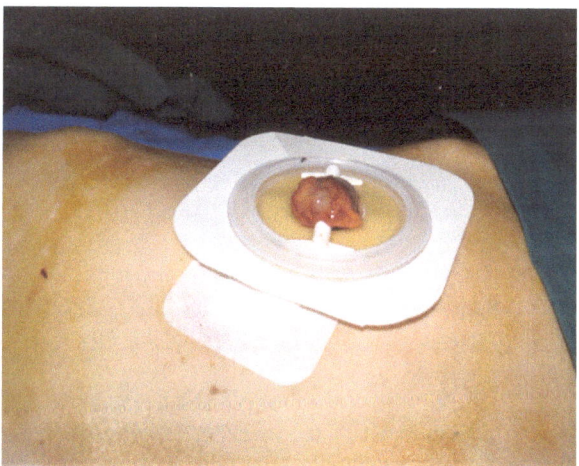

Fig. 7.6 Mixter's or Quenu's colostomy

fixed to the derma or the skin after its opening. The fecal material coming out from the stoma during a peristaltic wave is collected into a removable and disposable collection pouch (called also ostomy pouch or ostomy bag), which is then attached to the stoma. Normally it requires to be changed one to two times per day. Obese patients are at high risk of stoma-related complications because of the width of the subcutaneous fat requiring the exteriorization of a longer segment of the colon with its mesentery.

Ileostomy formation follows the same rules of the colostomy and is usually easier to perform (even laparoscopically) compared to a colostomy because of the natural mobility of this part of the bowel. The spout of the exteriorized ileum must be long enough to facilitate the position of the ileostomy bag in order to minimize the contact of the ileal content with the peristomal skin.

Lateral ileostomy is nowadays the preferred temporary bowel diversion in case of rectal cancer surgery [1, 2] also because it is easier to close when the restoration of the bowel continuity becomes possible. The position of the stoma is usually decided and marked on the skin before surgery using the same criteria of a colostomy (see above). In the obese patients, it is always the best type of intestinal diversion.

7.5   Pathophysiologic Implications

Left colostomy: left (distal) colostomy has minimal implication on the pathophysiology of nutrition in these patients, since the only organ bypassed is the rectum which normally is devoted to storage of the feces, minor effects on the final dehydration of the stools, and control of the defecation.

Nevertheless, the nutritionists should be aware of the profound changes in the microbiota induced by the mechanical bowel preparation and prolonged antibiotic therapy which normally occur in patients undergoing colorectal surgery and suggest pre/probiotic treatments.

The output through the stoma is formed stools, and the frequency of the bowel movement and consistency of the feces strongly depends on the preoperative bowel habit (patients with slow transit constipation remain constipated) and type of diet and drinking.

Coecostomy: from a pathophysiologic point of view, it must be considered like a terminal ileostomy (see later).

Ileostomy: the pathophysiologic consequences of an ileostomy strongly depend on the level of the ileostomy. In any case the role of the colon on the absorption of water, electrolytes, and biliary salts is lacking; therefore the stools will be fluid and irritating on the skin for the presence of biliary salts and residual activated pancreatic enzymes. Furthermore, the output from the stoma is frequent following the migrating motor complex of the small bowel that occur approximately every 90 min in fasting state. Furthermore, the passage of liquid stools often follows the oral feeding after few minutes.

This condition can cause dehydration and electrolyte imbalance leading to hypotension and kidney failure.

Hopefully this high output from the ileostomy tends to decrease within 2–3 months, and the stool becomes more consistent and dehydrated.

When the stoma involves the proximal part of the ileum or the jejunum, the pathophysiologic consequences may be devastating, and often the patients need to be hospitalized for a parenteral nutrition and rehydration.

In the long term, these patients can gain weight because the absorption of most of the nutrients is maintained, nevertheless they are predisposed to develop gallbladder stones because of the altered enterohepatic recirculation of the biliary salts absorbed in the cecum.

Approximately 8500 mL of fluid is added to the gastrointestinal tract daily, 1500 mL coming from the diet and 7000 mL coming from endogenous secretions from the salivary glands, stomach, small intestine, pancreas, and biliary tree. Of this amount, roughly 7000 mL is absorbed in the small intestine, leaving 1500 mL of

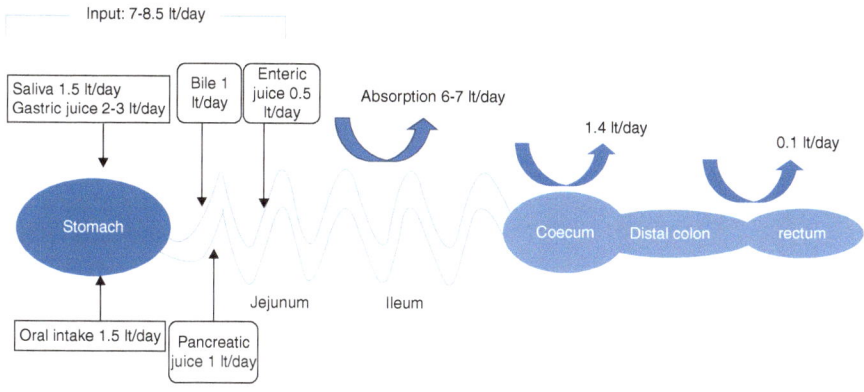

Fig. 7.7 Schematic fluid resorption in the gastrointestinal tract

fluid for the cecum to handle. In the cecum and proximal colon, a further 1400 mL is absorbed, leaving approximately 100 mL of fluid to be expelled with the solid fecal material in the stool. In contrast to the relatively leaky mucosal resistance of the jejunum and ileum, resistance in the colon becomes much tighter, thus preventing back diffusion of electrolytes and water. Associated with this prevention is an increase in the spontaneous membrane potential difference, which further restricts ionic movement into the lumen. The absorptive processes are different in the distinct regions of the gastrointestinal tract, with absorption of sodium, chloride, and short chain fatty acids (SCFA) predominantly in the [3] colon (Fig. 7.7).

7.5.1 Pathophysiology of Ileostomy and Nutritional Support in Patients with Ileostomy

The ileus is the most important part of the GI tract involved in the absorption of nutrients, such as carbohydrates, lipids, proteins, mineral salts, and vitamins, and in the reabsorption of the biliary salts. Therefore, following surgery on the ileus, it is important to know the length of the residual small bowel in order to estimate its ability to allow adequate absorption of nutrients and fluids: the higher the site of the ostomy, the higher the risk associated with nutritional impairment and malabsorption.

If the resection involves the jejunum, the possible deficiencies are iron, calcium, magnesium, and all the three macronutrients (protein, carbohydrate, and fat) [4, 5].

In case of distal resection of the ileus, the absorption of lipids, bile, and vitamin B12 might be significantly impaired [6]. Thereafter, the postoperative diet and possible need of supplements depend on the spared tract of gut, its physiological conditions, and the presence of associated diseases.

In the first months after the surgery, it is necessary to have a high caloric diet (40–60 kcal/kg of weight) with high protein content (1.5 g/kg of weight) in order to provide enough energy and nutrients to restore the muscle mass and promote early recovery. In case of obese patients, the diet should include a high percentage of proteins (22–25%) and a low quantity of carbohydrates, with a low glycemic index (do not overcome a glycemic load of 20 g per meal and 80 g daily). Lipid supplementation should be low in order to prevent steatorrhea and to decrease the overall calories during meals. High biological protein value with low fat presence should be preferred (low-fat meat, egg white, low-fat fishes, bresaola, low-fat ham, low-lactose and lactose-free cheese). The digestion of the lactose is often difficult even in patients without lactose intolerance, and therefore patients with intestinal stomas should prefer lactose-free foods.

In the presence of weight loss higher than 1 kg per week or irresponsive diarrhea (>600 g output per day), the need for parenteral nutrition for a limited period should be considered [7].

The control of the hydroelectrolytic balance is of pivotal importance in the presence of an ileostomy.

The estimated loss of NaCl through an ileostomy is about 30–40 mEql/die; however the small bowel adapts, over time, to reabsorb water, so the patient will decrease the stoma output from 1000/1200 mL/die to about 750 mL/die within 1–2 months from surgery; this adaption is easier and faster in distal ileostomy.

The most frequent cause of new hospitalization is dehydration with some cases of renal insufficiency. This risk is greater in patients under diuretic therapy for other reasons, including arterial hypertension. Symptoms of dehydration must be recognized as soon as possible (increased thirst, dry mouth, dry skin, decreased urine production, fatigue, shortness of breath, headache, dry eyes, and abdominal cramps) in order to be able to correct them promptly. To compensate for urinary and saline losses, the urine is more concentrated, predisposing to the calcium and uric acid precipitation and stone formation. Hypomagnesemia can also occur because of ileostomy loss, and therefore the support of magnesium-rich foods should be encouraged (almonds, sunflower seeds, fish, tofu, avocado). Vitamin B12 is mainly absorbed into the ileum, and therefore the risk of vitamin B12 deficiency must be considered in these patients leading to macrocytic anemias and possible repercussions on the folate cycle and hyperhomocysteinemia.

The passage of the chimo in the terminal part of the small intestine stimulates the release of entero-hormones like the Pyy and GLP2, with a modulating function on the appetite and gastric motility and secretion. This mechanism is known as ileal brake, decreases gastric emptying, increases pyloric contraction, slows postprandial transit, decreases pancreatic exocrine secretion, and modulates the appetite reducing food intake. As a consequence, resection of the terminal ileum can result in increased gastric motility and transit time. The ileal brake mechanism is completely lost in proximal ileostomies, while it is quite maintained in distal ileostomies.

In ostomy patients, particularly in the first 2–3 postoperative months, it is important to keep solid meals separate from liquids. Each meal must be composed by small portions of food and must have at least 3 h interval between meals.

To decrease the volume of feces and frequency of bowel movements, meals should be based on refined cereals in order to avoid the whole fiber intake, peel the fruit, and avoid the foods with the seeds (eggplant, zucchini, tomatoes, kiwi), which can speed up the transit. From the 6th to 8th°week, depending on the stoma output and individual response, the patient can switch to a semi-solid diet but with low amount of fiber and few fats.

It should be remembered that the small intestine secretes about 100 mmoles/L of sodium in the intestinal lumen; therefore it is necessary to consider as an integral part of the feeding also 90 mmol/L of sodium, which roughly corresponds to about 5 g of cooking salt, favoring an easy water absorption [8].

These 5 g of sodium chloride are in addition to the preoperative meals. Water intake too must be increased compared to the usual diet to prevent dehydration; the consumption of at least 500–700 mL of water per day should be added to normal preoperative water intake.

Isotonic solutions must be preferred, and drinking hypotonic solutions should be limited to no more than 500 mL per day.

Examples of suggested drinks can be a mix of orange juice, about half a liter, one teaspoon of sodium chloride, one teaspoon of baking soda, and sugar, which increases the absorption of sodium and should be added depending on the patient's glucose tolerance, and water up to a liter to be drunk away from meals [9].

Patient with jejunostomy may exhibit supervisory changes due to hyperammonemia, since the residual intestine does not produce an appropriate amount of citrulline to detoxify the ammonium, through the urea cycle. Prescription of arginine assumption may be able to decrease the serum levels of ammonium, thanks to the activation of the urea cycle.

Calcium oxalate lithiasis is also common in patients with ileostomy (in the colostomy patients, instead, uric acid lithiasis is favored) due to a decrease in diuresis. Alkalization of the urine (pH > 6.5) can help preventing urolithiasis. This may be achieved using negative PRAL foods (potential renal acid load) (vegetables, beans, barbet, milk, bananas).

Patients presenting steatorrhea are at risk of developing hyperoxaluria due to increased permeability of the oxalate. In this case the aim of the diet is to reduce fatty acids and oxalates and/or increase calcium intake to bind oxalates in the intestinal lumen and in urine. Foods rich in oxalate are rhubarb, spinach, beets, cocoa and chocolate, sweet potatoes, strawberries, celery, and peanuts, which should be removed or reduced to a minimum.

Magnesium citrate can also be prescribed to decrease the risk of urolithiasis [10].

Patient with ileostomy should chew the food completely; chewing allows a better digestion and a more efficient absorption of nutrients, decreasing the possibility of obstruction of the stoma.

The evening meal must be taken at least 3 h before going to bed and must be of small quantity in order to decrease the stoma output during the night limiting the necessity to change the bag during the night [11].

Insoluble fiber, like bran, makes the stool more voluminous and accelerates the transit; it causes a greater loss of bile and a lower absorption of cholesterol; therefore it can be recommended when the patient has a hypercholesterolemia.

While soluble fibers, such as pectin, decrease gastric emptying and transit time, it slows the absorption of carbohydrates and can be therefore recommended in patients with diabetes [12].

In the first weeks after ileostomy, the fibers should be reduced to a minimum in order to decrease the fecal volume, and then it must be reintroduced progressively by choosing the right fiber-rich foods in relation to the intestinal transit and stool consistency.

The absorption of trace elements may be decreased by 10–20% in phytate-rich diets [13]. Phytate-rich foods include whole grains, legumes, soy, peanuts, sesame, and cocoa powder, which should be eliminated from the diet of patients bearing proximal ileostomies.

7.5.2   Nutritional Support in Patients with Colostomies

One of the main aims of the colon is the resorption of the water contained in the feces that leave the ileum. It is estimated that only 40–400 mL of fluids are found in the feces out of 800–1800 mL that enter the colon daily. In addition, it also has the function of reabsorbing electrolytes, especially sodium, vitamin B12, and vitamin K, particularly in the right colon.

Consequently, in the left colon ostomies, the colonic functions are almost completely preserved. Therefore, minimal nutritional advices are required in the daily diet.

In order to have regular bowel movements, these patient should drink adequate quantity of water (1500–2000 mL) and high fiber diet; the amount of fibers to be ingested, however, should be adjusted *case by case* in order to prevent the passage of high volume stools requiring frequent changes of the bag.

Soluble fibers could be added in the diet in order to reduce the absorption of carbohydrates and to provide nutrients to the microbiota. Soluble fibers are found mainly in fruit eaten with peel (the richest part of pectin is the white layer below the peel, called albedo, then removing the peel often eliminates pectin), in legumes, in chicory, and in oily fruit. Insoluble fiber-rich foods are present in whole foods, in the outer part of legumes, and in fibrous vegetables.

The choice of one type or the other type of fibers will depend on the occurrence of concomitant diseases (diabetes, dyslipidemia, use of anticoagulants) and the transit time, by the state of hydration of the patient and the amount of oil assumed. Therefore, sometimes the introduction of the fibers alone cannot correct constipation.

Stomas performed on the cecum or right colon should be managed like patients with distal ileostomies. However, this type of colostomy is going to be abandoned in favor of a terminal ileostomy.

In patients with documented lactose intolerance, the consumption of containing lactose foods like milk should be limited or better replaced with "milk" of vegetable origin (soy, rice, oats, almonds or other), while soy derivatives can be used instead of cheeses.

In any case, it is necessary to suggest the patient to make a good chewing of all the foods, in order to make the nutrients more easily absorbable, while the fibers, both insoluble and soluble, are reduced to pulp and do not increase gas production or alterations of the bowel motility.

Water intake should be preferred between meals, in small and frequent portions, about 80–120 mL for 12–10 times a day; this quantity should be increased if the feces become fluid and abundant.

Another problem related to the nutritional support in patients with a colostomy is the occurrence of abnormal meteorism with consequent inflate of the stoma bag. In these cases, chewing gum, carbonated drinks, and foods that induce the production of gases such as brassicas (cabbage, cauliflower, turnips, broccoli, etc.) should be avoided.

If consumed with a complementary food such as cereals, legumes are a good source of high biological value proteins and contain both soluble and insoluble fibers, however, the indigestible sugars contained into the legumes (raffinose, stachyose, and verbascose) reach unaltered the terminal part of the small intestine and, if present, the large intestine where they are fermented by the local bacterial flora with gas production. A prolonged soaking, at least 48 h, allows the germination of legumes with the use by the same legumes of their indigestible sugars; therefore this would lead to a lower presence in cooked ones. In addition, germination makes the minerals and vitamins present in the legume more available for absorption [14, 15].

References

1. Williams NS, Nasmyth DG, Jones D, Smith AH. De-functioning stomas: a prospective controlled trial comparing loop ileostomy with loop transverse colostomy. Br J Surg. 1986;73(7):566–70.
2. Gooszen AW, Geelkerken RH, Hermans J, Lagaay MB, Gooszen HG. Temporary decompression after colorectal surgery: randomized comparison of loop ileostomy and loop colostomy. Br J Surg. 1998;85(1):76–9.
3. Sellin J, Shelat H. Short-chain fatty acid (SCFA) volume regulation in proximal and distal rabbit colon is different. J Membr Biol. 1996;150(1):83–8.
4. Tsao SK, Baker M, Nightingale JM. High-output stoma after small-bowel resections for Crohn's disease. Nat Clin Pract Gastroenterol Hepatol. 2005;2:604–8.
5. Zarkovi M, Milievi M. Enteral nutrition in patients with ileostomies and jejunostomies. Acta Chir Iugosl. 1995;42:17–20.
6. Tilg H. Short bowel syndrome: searching for the proper diet. Eur J Gastroenterol Hepatol. 2008;20(11):1061–3. https://doi.org/10.1097/MEG.0b013e3283040cc9.
7. Seetharam P, Rodrigues G. Short bowel syndrome: a review of management options. Saudi J Gastroenterol. 2011;17(4):229–35. https://doi.org/10.4103/1319-3767.82573.
8. Van Gossum A, Cabre E, Hébuterne X, Jeppesen P, Krznaric Z, Messing B, Powell-Tuck J, Staun M, Nightingale J. ESPEN guidelines on parenteral nutrition: gastroenterology. Clin Nutr. 2009;28(4):415–27. https://doi.org/10.1016/j.clnu.2009.04.022.
9. Nightingale J, Woodward JM. Small bowel and nutrition committee of the British society of gastroenterology. Guidelines for management of patients with a short bowel. Gut. 2006;55(Suppl 4):iv1–12.
10. Worcester EM. Stones from bowel disease. Endocrinol Metab Clin N Am. 2002;31(4):979–99.
11. Riobó P, Sánchez Vilar O, Burgos R, Sanz A. Colectomy management. Nutr Hosp. 2007;22(Suppl 2):135–44. Review

12. Ellegård L, Andersson H. Oat bran rapidly increases bile acid excretion and bile acid synthesis: an ileostomy study. Eur J Clin Nutr. 2007;61(8):938–45.
13. Agte V, Jahagirdar M, Chiplonkar S. Apparent absorption of eight micronutrients and phytic acid from vegetarian meals in ileostomized human volunteers. Nutrition. 2005;21(6):678–85.
14. El-Adawy TA, Rahma EH, El-Bedawey AA, El-Beltagy AE. Nutritional potential and functional properties of germinated mung bean, pea and lentil seeds. Plant Foods Hum Nutr. 2003;58:1–13.
15. Marton M, Mandoki ZS, Csapo-Kiss CJ. The role of sprouts in human nutrition. A review. Acta Univ Sapientiae Aliment. 2010;3:81–117.

Nutritional Support After Surgery of the Liver

Riccardo Memeo and Anna D'Eugenio

8.1 Introduction

Liver resection remains the only curative treatment for hepatic malignancies. Many progresses have been recently done in the field of chemotherapy that, in association with surgery, could guarantee the best survival for patients with liver malignancies including primitive cancer (hepatocellular carcinoma, cholangiocarcinoma) and metastatic cancer (colorectal liver metastasis, breast cancer). The main concept of resectability in liver surgery is associated to the concept of leaving a sufficient quote of liver parenchyma to guarantee a sufficient liver function which is mandatory to reduce the risk of postoperative liver failure. Recently, several improvements in the perioperative management of surgical patient have been introduced. For example, prediction of future remnant liver could be calculated with 3D reconstruction software, as well as quality of liver function with the ICG test. Minimally invasive surgery (robotic and laparoscopic) has also contributed to minimize the surgical trauma. Furthermore, adhesion to the ERAS (enhanced recovery after surgery) protocols has contributed to reduce postoperative in hospital stay.

8.2 Anatomy of the Liver

Morphologically, the liver is divided in the right and left lobe, with an ideal line through the falciform ligament. The most important contribution to the surgical anatomy of the liver comes from Couinaud in 1957, who divided the liver in eight anatomical and functional segments, who could be separately resected without

R. Memeo (✉)
Department of Emergency and Organ Transplantation, University of Bari, Bari, Italy
e-mail: info@drmemeoriccardo.com

A. D'Eugenio
Consultant Nutritionist, Ambulatorio di Medicina Integrata, G. Bernabei Hospital, Ortona, Italy

compromise the function of other segments. This distribution is based upon the distribution of portal pedicle and hepatic veins. Right, middle and left hepatic veins divide the liver in four sectors, and each of them receives independent portal supply. This allows the division of the liver in eight segments and the consequential classification of liver resection. Liver resection however may follow anatomical and non-anatomical planes. Anatomical hepatectomies resect a portion of the liver and its vascular supply. Non-anatomical hepatectomies resect a portion of parenchyma, not limited by anatomical landmarks, for example, a lesion who is among segments 5 and 6 could be resected without resecting the totality of segments 5 and 6 (also known as wedge resection). In function of number of resected segments, we defined major hepatectomies if three or more segments are resected. Minor hepatectomies are defined if two or less segments are resected. Most common operations are left and right hepatectomy, bisegmentectomy, segmentectomy and wedge resection.

8.3   Evaluation of Liver Function

Main risks related to liver resection are hepatic insufficiency and failure [1]. This risk is increased in case of an excessively large amount of the hepatic parenchyma liver resection [2]. For that reason, the preoperative risk assessment is a fundamental process before liver resection. In case of liver resection, a preoperative quantitative evaluation of the percentage of residual hepatic parenchyma [3] and a qualitative [4] evaluation (functional liver reserve) should be performed. According to the quality of liver parenchyma, we should preserve at least 25–50% of the remnant liver, especially in case of cirrhotic, steatosis and post-chemotherapy liver. This evaluation can be achieved preoperatively with 3D reconstruction models, based on CT scan or MRI reconstruction.

For the qualitative evaluation, the main test is the retention rate of indocyanine green at 15 min. Other items evaluated before liver resection is portal hypertension [5] and the Child-Pugh classification [6]. This classification is used to assess the prognosis of chronic liver disease, mainly in cirrhotic patients (Fig. 8.1). It is based on the analysis of five items (total bilirubin, serum albumin, prothrombin time or

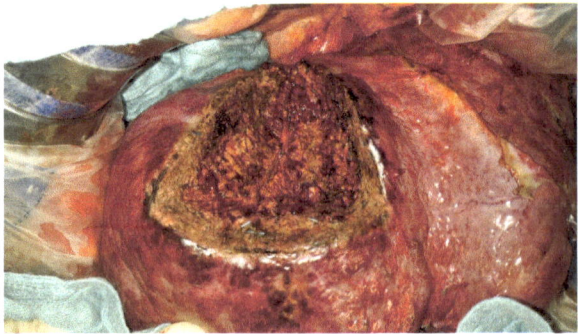

Fig. 8.1 Liver resection for hepatocellular carcinoma on cirrhotic liver

INR, ascites and hepatic encephalopathy) and identifies three classes of patients according to the cumulative score.

Most recently, a new concept in liver resection has been introduced, to minimize the risk of postoperative liver failure and to improve the possibility to treat recurrent pathology: the parenchymal-sparing hepatectomies. This strategy consists in performing multiple resections instead of a single major hepatectomy, whenever possible. It has been demonstrated that this strategy is feasible and safe, especially in liver resection for colorectal liver metastasis.

8.4   Liver Resection

With a 50% 5-year overall survival, liver resection is considered the mainstay of therapy in patient with preserved liver function. Many progress have been done in recent era to improve results of liver resection. Better patient selection and preoperative studies, associated with improvement of surgical tools and techniques, such as laparoscopic [7] and robotic surgery [8] (Fig. 8.2a, b), have improved postoperative outcome. Preoperative assessment of the patient plays a key role.

First of all, a CT scan of the abdomen and thorax is mandatory, to exclude contraindication for surgery such as major parenchymal involvement or distal metastasis. The role of the CT is at the same time to establish a correct diagnosis and to show the relationship of nodules with vascular and biliary structures. In case of major resection, it is mandatory to calculate, based on CT 3d reconstruction, the amount of theoretical future remnant liver (FRL) [9] who corresponds at the quantity of liver who should remain after surgery and who should be sufficient to guarantee a normal liver function. In case of insufficient FRL, portal vein embolization [10] (selective occlusion of monolateral portal flow to obtain contralateral hypertrophy of the liver) could be useful to increase FRL. In case of major resection, at least 40% of FRL should be preserved in cirrhotic patient.

The most important aspect in liver resection is the identification of appropriate candidate who could underwent liver resection. A correct assessment of patient

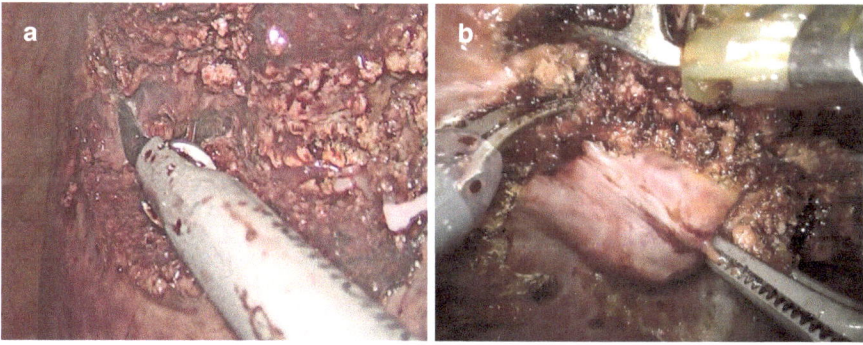

Fig. 8.2 (**a**, **b**) Laparoscopic (left) and robotic (right) liver resection

general status and liver function must be performed to reduce at minimum the guarantee of an uneventful postoperative course. Nowadays, it is possible to treat bilobar liver metastasis, and strategies in association with chemotherapy could allow treatment of previously unresectable liver lesions. Two-stage hepatectomies and ALLPS permit today the treatment of bilobar liver disease with good postoperative mortality and oncological results.

8.4.1 Surgical Techniques for Liver Resection

The aim of liver resection is to offer the best treatment, with adequate resection margin [11, 12]. Generally, a 1 cm (better 2 cm) resection margin tumour free (Fig. 8.3a, b) should be maintained in order to remove satellite nodules eventually present preventing an early recurrence of the disease. For the same reason, anatomical resection is preferred to nonanatomical resection [13] due to intrahepatic cancer diffusion following the portal vein. In most cases, an associated limphectomy is mandatory to improve the oncological outcome.

Liver resection needs an initial intraoperative ultrasound, in order to identify liver lesions and anatomical relation between cancer and vascular/biliary structures. Once the criteria for a safe resection are identified, liver resection could be performed using different techniques and devices, in order to reduce blood loss and to perform an easy hepatectomy [14]. In most of liver resection, a tape is passed around the round ligament in order to have the possibility to clamping the inflow (Pringle manoeuvre) of the liver and to control an uncontrolled intraoperative bleeding, even if the duration of pedicle clamping must be limited to few minutes.

In last 20 years, the improvement and diffusion of laparoscopic liver surgery, associated with the development of new surgical tools, have made easier the minimally invasive resection of the liver. Apart from the advantage of minimally invasive

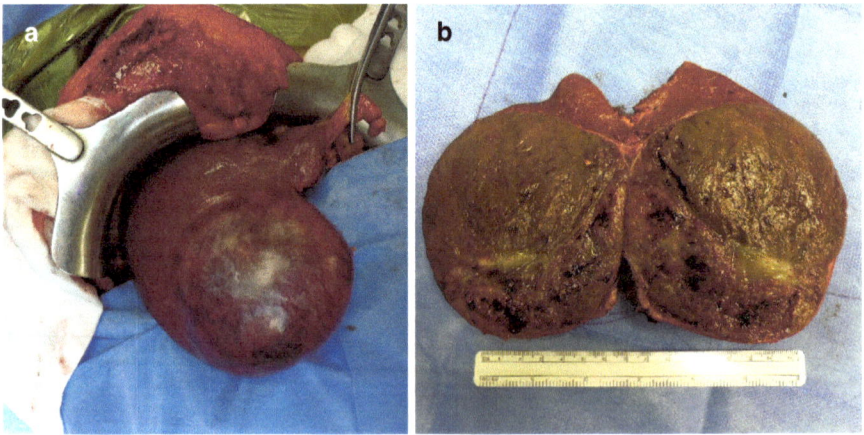

Fig. 8.3 (**a**, **b**) Operative view of hepatocellular carcinoma and specimen after resection

access on postoperative pain, laparoscopic liver surgery has been demonstrated to reduce intraoperative bleeding, allow faster recovery and have same short-term and long-term oncological results [15]. It is possible to associate liver resection to radio-frequency ablation.

8.5   Pathophysiology of Nutrition After Liver Surgery

The regenerative capacity of the liver (about 50 days for complete regeneration) makes nutrition of fundamental importance after major resection surgery [16, 17]. Helping the organ to recover structurally and functionally, within the limits of each individual's possibility, allows an improvement in the metabolisms and function of removing the toxins that enter the portal blood [18, 19]. In liver diseases there are symptoms that pertain to the metabolism of glucose (the liver carries out glycogen-synthesis, glycogenolysis, gluconeogenesis), of fatty acids and cholesterol, of the neo-synthesis and of protein catabolism but also the detox processes [20].

Individual variability regarding this function is very high; therefore, after major liver resection, it is necessary to reduce the introduction of toxic substances to a minimum [21]. Even though the liver can double the surface of the smooth endoplasmic reticule, increasing the quantity of enzymes needed for the toxification and removing a large part of the organ reduce the functions of sulphur and glucuronide conjugation that allows the biliary elimination of toxins.

8.5.1   Nutrition Advices After Liver Surgery

It is very important:

- To avoid the toxic substances linked to production methods of foodstuffs (residue of pesticides, agrochemicals, weed-killers)
- To avoid food additives (colouring, preservers, flavouring)
- To avoid certain cooking methods which develop toxic substances such as acrylamide, acrolein, heterocyclic aromatic ammines

Consequently, it is important to use foodstuffs from biodynamic or biological agriculture [22], freshly prepared so as to avoid preservers and cooked so as not to produce substances toxic for the organism (avoid frying, grilling and barbecuing) [23, 24].

In addition, salt should be used in small quantities, as there is always a slight/moderate hypoalbuminemia.

Assessment of symptoms helps understand which functions are most affected. Digestive disorders, a bitter taste in the mouth, difficult digestion, a sense of heaviness after eating, anorexia, meteorism and irregular evacuation are always present, as also are coagulation disorders, since the liver produces many of the specific factors (prothrombin, fibrinogen, factors V, VII, IX and X) and destroys the plasmatic activator of plasminogen.

Very often anorexia must lead to the choice of a tasty diet, where flavours and attractive aromas increase the appetite for food and contrast the tendency to undereat. Malnutrition is dangerous because it worsens the prognosis of the disease [25]. To avoid getting trapped in food ideologies, it is as well to remember that there are lifestyles able to improve life but that a nutrition that is miraculous for cancer does not exist [26].

In the days following surgery, the patient may be constipated, a side effect of many pain-killers. It is a good idea to eat five small meals as the digestive potential will be reduced because of the removal of a considerable part of the liver [27]. Meals will have to ensure a gradual release of sugars in order not to come up against moments of hypoglycaemia. Despite this, the quantity of fibre and cooked leafy vegetables should be reduced. Creamed vegetable soups should be preferred but also steamed, boiled or stewed courgettes and carrots. Cereals, semi-wholemeal pasta cooked in vegetable broth or with tomato sauce or creamed vegetables and limited amounts of meat and fish, are meals able to provide energy without negative side effects [28].

Prostration, alteration of the faeces with a greater tendency to diarrhoea, weight loss, lack of appetite and a sense of satiety after meals are the symptoms that frequently occur in the following weeks and which can be contrasted by using aromatic herbs and spices (rosemary, sage, thyme, parsley, basil, curcuma, ginger but also lemon juice, wine and apple vinegar) and varying cooking methods. It is better not to use only boiling and steaming but to introduce stewing and baking in foil. The various digestive stimuli will increase the production of gastric juices and therefore of digestives enzymes [29, 30].

When the disorders linked to the operation have passed and the will to eat returns more or less to normal, the following diet can be hypothesized:

- Breakfast: chicory or barley coffee or tea or green tea, toasted semi-wholemeal bread with jam or honey, plain yogurt, fruit
- Mid-morning: fresh vegetable and fruit juice, toasted semi-wholemeal bread seasoned with oil and salt or with egg white
- Lunch: pasta or rice or other cereals in grains (spelt barley, oats, quinoa, millet), white meat (an average of 120 g) or fish (an average of 170 g), raw vegetables
- Mid-afternoon: fresh vegetable or fruit juice or fresh fruit
- Dinner: creamed vegetable soup, rice or other cereals in grains seasoned in various ways or mixed with sieved pulse, raw vegetables

In chronic liver disease, raw vegetables prove to be more useful than cooked vegetables (for their abundance of vegetation liquid); however, to improve the amount of vitamin K, important for coagulation, which is always affected in cases of liver disease, it is better to eat a vegetable which contains this vitamin at least every other day and as it is a liposoluble vitamin, to cook it with extra virgin olive oil (spinach, cabbage, carrots, etc.) [27, 31–33].

Besides the amount assumed in food, the most important quantity of vitamin K is produced by intestinal bacteria (above all Escherichia), a condition enhanced by a good microbiota [34]. This represents the variable component of the human genome, and being variable it can be modified by nutrition. The bacteria grow on the swallowed foods, so eating vegetable products and fibre favour the growth of the anti-inflammatory kind, while the pro-inflammatory species are facilitated by a diet rich in protein. Moreover, the intestinal bacteria, using the fibre which we swallow, produce short-chain fatty acids which are absorbed into circulation regulating the immune system and reducing inflammation, with an improvement in the course of neoplastic diseases [35, 36].

Scientific literature does not agree about the amount of necessary protein which some authors advice high, others lower. A diet containing 1.2–1.5 g of protein per kilo seems to be adequate, with 5–6 g of carbohydrates per kilo. Fruits and vegetables can be eaten freely. As well as vitamin K, very important for the activation of prothrombin, the diet should try to introduce adequate amounts of thiamine (present in cereals, pulse and meat), ascorbic acid (citrus fruits) and vitamin B12 (animal derivatives). These vitamins are mostly present in the vegetable realm and in particular those of the B group are found in leafy vegetables; as they are hydrosoluble vitamins, they are assimilated better if the vegetables are eaten raw.

It is very important to avoid drinking alcohol which is metabolized and disposed off by the liver in its non-toxic components. Alcohol consumption worsens the prognosis and speeds up the advancement of the disease.

Attention should be paid to fructose. This monosaccharide is found freely, together with glucose and galactose, in honey and fruit, but it also found added to the syrups and in the sweeteners used mainly in the soft drinks industry and in various processed foods [37]. Fructose is mainly metabolised in the liver. This process of synthesis produces energy for the body but also other derivatives such as uric acid. If the amount of fructose swallowed systematically is excessive, the metabolic pathway is altered, and too much uric acid is produced which, if not well disposed of, triggers inflammatory processes of the liver cells, resistance to insulin and oxidative stress [38]. The fructose present in fruit does not cause the same effects probably because of the presence of vitamin C which increases the renal excretion of uric acid.

Use only extra virgin olive oil for seasoning, which thanks to its tocopherols, to the phenolic and triterpene compounds, carries out a powerful anti-inflammatory action and helps to maintain the sugar level stable.

Two vegetables deserve special attention, the artichoke and the edible thistle. Thanks to cynarine, a derivative of caffeic acids, with a choleretic and cholagogue action and to inulin, an oligosaccharide which acts as a probiotic, the artichoke improves the hepatocyte functioning [33]. No less useful is the edible thistle which thanks to its silymarin content seems able to protect the liver cells from oxidation damage and to stimulate the synthesis of hepatic proteins. If these foods were poorly tolerated at an intestinal level, for the development of meteorism, they can be taken in the form of phytotherapy supplements.

8.6   Pathophysiology of Nutrition After Major Resection in Patients with Cirrhosis

When major liver resection is carried out in a patient affected by liver cirrhosis, there is often a deterioration of the general condition [39]. A medical history of liver failure can be of a slight, medium or severe degree according to the functions involved and to the extent of the impairment. The nutritional assessment which follows refers to the first two forms.

The distorted lobular structure, the result of necrotic phenomena with the replacement of necrotic liver parenchyma with fibrous tissue, alters the capacity of the hepatocytes to acquire portal blood. The symptoms that the patient presents are linked mostly to the presence of systemic portal shunts, so that an important amount of blood does not pass through the sinusoids and the liver cells do not succeed in carrying out a plasma exchange.

Compared to the patient with major liver resection who does not present a medical history of cirrhosis, jaundice may easily appear, an expression of alterations in the cellular metabolism of bilirubin and its excretion. Clinical signs are itching and yellow skin colour. The indices of cytolysis are more or less altered but in this form ammonia and pressure of the portal vein begin to increase.

Portal hypertension reduces liver cell exposition to portal vein glucose and leads to an altered glucose tolerance. The increase of glucagon brings about an increase of gluconeogenesis, so the cirrhotic liver does not succeed in storing glucose even though it preserves the other functions. Overtime, however, gluconeogenesis may fail and cause negative hypoglycaemic crises. The worst complications for the cirrhotic patient (bacterial peritonitis, cirrhotic encephalopathy, intestinal perforation, ascites) are caused by portal hypertension which detaches the layer of intestinal mucus and loosens the tight junctions, and so bacteria go into the bloodstream. It is difficult to locate these bacteria in the bloodstream because, being anaerobic extremophiles, they die as soon as they are in contact with blood. But their translocation fragments, the toxic fragments, enter the bloodstream creating a condition of dangerous inflammation. The study of the intestinal microbiota is of primary importance, and perhaps the newly acquired knowledge will be fundamental for the management of portal-systemic encephalopathy [40].

The most important symptoms that indicate the presence of a cirrhotic medical history and that worsen the post-operative course are a more marked anorexia, weight-loss and the loss of muscle mass.

The anomalies of glycaemic, lipidic and protein metabolism make the nutrition of the patient who has undergone surgery for hepatic carcinoma with a cirrhotic liver very complicated, and, when the condition evolves towards severe liver failure with portal hypertension and encephalopathy, it is no longer sufficient on its own [41].

When there is liver failure, there is an increase of aromatic amino acids (phenylalanine, tyrosine, tryptophan, methionine, glutamate, aspartate and histidine) to the detriment of the branched ones (leucine, isoleucine and valine). The increase of peripheric consumption linked to the empidid production and to the use of energy sub-layers (glucose, fatty acids and ketonic bodies) explains the reduction of the latter.

8.6.1   Dietary/Nutritional Advice After Major Resection in Patients with Cirrhosis

It is important for the diet to ensure a certain amount of carbohydrates with a low glycaemic index at every meal, so as to avoid glycaemic anomalies. Carbohydrates in grains (spelt, pearl barley, wholemeal rice, oats) are more useful than pasta and bread. Sweet desserts are forbidden, and the quantity of fruit must be controlled, only 100–150 g mid-morning and 100–150 g mid-afternoon. Fresh vegetable juices (in this case with very little fruit) are still very useful, providing water, mineral salts and enzymes, which improve digestion.

If the cirrhosis is compensated, protein consumption can remain that of a healthy person, about 1.2 g/k. If the clinical conditions are deteriorating, the amount should be progressively reduced to as low as 0.5 g/k [41]. In decompensated cirrhosis, in fact, although the amount of protein is fundamental to maintain an adequate protein synthesis and liver regeneration and also to support the immune system, the serious anomalies of the metabolism of nitrogen impose rigid dietary restrictions. The egg remains the best tolerated protein source, its white is rich in branched amino acids, and its renal involvement is very slight. If eggs are steamed or boiled, they do not create digestive problems, and in the author's experience, they can be used once or twice a week to great advantage. In the case of bilirubin values above 2 mg/mL, only egg white can be used, avoiding the yoke, which is too fatty.

In case of the advancement of liver failure (presence of portal hypertension, oesophageal varices, hyper-ammonia, ascites), a useful nutritional strategy is that of concentrating proteins in the earlier part of the day, avoiding them in the evening meal. This will lead to a better renal management both of nitrogenous waste and of acid equivalents and consequently to a clinical improvement.

References

1. Rahnemai-Azar AA, Cloyd JM, Weber SM, Dillhoff M, Schmidt C, Winslow ER, et al. Update on liver failure following hepatic resection: strategies for prediction and avoidance of post-operative liver insufficiency. J Clin Transl Hepatol. 2018;6(1):1–8.
2. Chan J, Perini M, Fink M, Nikfarjam M. The outcomes of central hepatectomy versus extended hepatectomy: a systematic review and meta-analysis. HPB (Oxford). 2018;20:487–96.
3. Cieslak KP, Huisman F, Bais T, Bennink RJ, van Lienden KP, Verheij J, et al. Future remnant liver function as predictive factor for the hypertrophy response after portal vein embolization. Surgery. 2017;162(1):37–47.
4. de Baere T, Teriitehau C, Deschamps F, Catherine L, Rao P, Hakime A, et al. Predictive factors for hypertrophy of the future remnant liver after selective portal vein embolization. Ann Surg Oncol. 2010;17(8):2081–9.
5. Rhaiem R, Piardi T, Chetboun M, Pessaux P, Lestra T, Memeo R, et al. Portal inflow modulation by somatostatin after major liver resection. Ann Surg. 2018;267:e101 3. http://www.ncbi.nlm.nih.gov/pubmed/29189385
6. Okajima C, Arii S, Tanaka S, Matsumura S, Ban D, Ochiai T, et al. Prognostic role of child-Pugh score 5 and 6 in hepatocellular carcinoma patients who underwent curative hepatic resection. Am J Surg. 2015;209(1):199–205.

7. Wakabayashi G, Cherqui D, Geller DA, Buell JF, Kaneko H, Han HS, et al. Recommendations for laparoscopic liver resection: a report from the second international consensus conference held in Morioka. Ann Surg. 2015;261(4):619–29.

8. Giulianotti PC, Bianco FM, Daskalaki D, Gonzalez-Ciccarelli LF, Kim J, Benedetti E. Robotic liver surgery: technical aspects and review of the literature. Hepatobil Surg Nutr. 2016;5(4):311–21.

9. Cieslak KP, Runge JH, Heger M, Stoker J, Bennink RJ, van Gulik TM. New perspectives in the assessment of future remnant liver. Dig Surg. 2014;31(4–5):255–68.

10. Memeo R, De Blasi V, Adam R, Goéré D, Azoulay D, Ayav A, et al. Parenchymal-sparing hepatectomies (PSH) for bilobar colorectal liver metastases are associated with a lower morbidity and similar oncological results: a propensity score matching analysis. HPB (Oxford). 2016;18:781–90.

11. Pauli EM, Staveley-O'Carroll KF, Brock MV, Efron DT, Efron G. A handy tool to teach segmental liver anatomy to surgical trainees. Arch Surg. 2012;147(8):692–3.

12. Memeo R, De'Angelis N, Compagnon P, Salloum C, Cherqui D, Laurent A, et al. Laparoscopic vs. open liver resection for hepatocellular carcinoma of cirrhotic liver: a case-control study. World J Surg. 2014;38(11):2919–26.

13. Huang X, Lu S. A meta-analysis comparing the effect of anatomical resection vs. non-anatomical resection on the long-term outcomes for patients undergoing hepatic resection for hepatocellular carcinoma. HPB (Oxford). 2017;19(10):843–9.

14. Appéré F, Piardi T, Memeo R, Lardière-Deguelte S, Chetboun M, Sommacale D, et al. Comparative study with propensity score matching analysis of two different methods of transection during hemi-right hepatectomy: ultracision harmonic scalpel versus cavitron ultrasonic surgical aspirator. Surg Innov. 2017;24(5):499–508.

15. Sotiropoulos GC, Prodromidou A, Kostakis ID, Machairas N. Meta-analysis of laparoscopic vs open liver resection for hepatocellular carcinoma. Updat Surg. 2018;9(3):291–311.

16. Nagasue N, et al. Human liver regeneration after major hepatic resection. A study of normal liver and liver with chronic hepatitis and cirrhosis. Ann Surg. 1987;206:30–9.

17. Michalopoulos GK, De Frences MC. Liver regeneration. Science. 1997;276:60.

18. Harimoto N, Hoshino H, Muranushi R, Hagiwara K, Yamanaka T, Tsukagoshi M, Igarashi T, Watanabe A, Kubo N, Araki K, Shirabe K. Skeletal muscle volume and intramuscular adipose tissue are prognostic predictors of postoperative complications after hepatic resection. Anticancer Res. 2018;38(8):4933–9.

19. Arias JM, et al., editors. The liver: biology and pathology. 3rd ed. New York: Raven Press; 1994.

20. Tajiri K, et al. Liver physiology and liver diseases in the elderly. World J Gastroenterol. 2013;19:8459–67.

21. Newberne P, Corner MW. Food additives and contaminants: an update. Cancer. 1986;58:1851–62.

22. Guzzon R, et al. Evaluation of the oenological suitability of grapes grown using biodynamic agriculture: the case of a bad vintage. J Appl Microbiol. 2016;120:355–65.

23. Oates L, Cohen M, Braun L, Schembri A, Taskova R. Reduction in urinary organophosphate pesticide metabolites in adults after a week-long organic diet. Environ Res. 2014;132:105–11.

24. Göen T, Schmidt L, Lichtensteiger W, Schlumpf M. Efficiency control of dietary pesticide intake reduction by human biomonitoring. Int J Hyg Environ Health. 2017;220(2 Pt A):254–60.

25. Okuno M, Goudmard C, Kopetz S, Vega EA, Joechle K, Mizuno T, Tzeng CD, Chun YS, Lee JE, Vauthey JN, Aloia TA, Conrad C. Loss of muscle mass during preoperative chemotherapy as a prognosticator for poor survival in patients with colorectal liver metastases. Surgery. 2019;165:329–36.

26. Baudry J, Méjean C, Allès B, Péneau S, Touvier M, Hercberg S, Lairon D, Galan P, Kesse-Guyot E. Contribution of organic food to the diet in a large sample of French adults (the NutriNet-Santé Cohort Study). Nutrients. 2015;7(10):8615–32.

27. Zhou Y, Li Y, Zhou T, Zheng J, Li S, Li HB. Dietary natural products for prevention and treatment of liver cancer. Nutrients. 2016;8(3):156.

28. Goetzke B, Nitzko S, Spiller A. Consumption of organic and functional food. A matter of well-being and health? Appetite. 2014;77:94–103.
29. Kim JY, Kwon O. Culinary plants and their potential impact on metabolic overload. Ann N Y Acad Sci. 2011;1229:133–9.
30. Tiwari AK. Revisiting "vegetables" to combat modern epidemic of imbalanced glucose homeostasis. Pharmacogn Mag. 2014;10(Suppl 2):S207–13.
31. Sung B, Prasad S, Yadav VR, Aggarwal BB. Cancer cell signalling pathways targeted by spice-derived nutraceuticals. Nutr Cancer. 2012;64(2):173–97.
32. Kannappan R, Gupta SC, Kim JH, Reuter S, Aggarwal BB. Neuroprotection by spice-derived nutraceuticals: you are what you eat! Mol Neurobiol. 2011;44(2):142–59.
33. Tapsell LC, Hemphill I, Cobiac L, Patch CS, Sullivan DR, Fenech M, Roodenrys S, Keogh JB, Clifton PM, Williams PG, Fazio VA, Inge KE. Health benefits of herbs and spices: the past, the present, the future. Med J Aust. 2006;185(4 Suppl):S4–24.
34. Brenner DA, Paik YH, Schnabl B. Role of gut microbiota in liver disease. J Clin Gastroenterol. 2015;49(Suppl 1):S25–7,
35. de Faria Ghetti F, Oliveira DG, de Oliveira JM, de Castro Ferreira LEVV, Cesar DE, Moreira APB. Influence of gut microbiota on the development and progression of nonalcoholic steato-hepatitis. Eur J Nutr. 2018;57(3):861–76. https://doi.org/10.1007/s00394-017-1524-x.
36. Acharya C, Bajaj JS. Gut microbiota and complications of liver disease. Gastroenterol Clin N Am. 2017;46(1):155–69. https://doi.org/10.1016/j.gtc.2016.09.013.
37. Jamnik J, Rehman S, Mejia SB, de Souza RJ, Khan TA, Leiter LA, Wolever TMS, Kendall CWC, Jenkins DJA, Sievenpiper JL. Fructose intake and risk of gout and hyperuricemia: a systematic review and meta-analysis of prospective cohort studies. BMJ Open. 2016;6(10):e013191. https://doi.org/10.1136/bmjopen-2016-013191.
38. ter Horst KW, Serlie MJ, et al. Fructose consumption, lipogenesis, and non-alcoholic fatty liver disease. Nutrients. 2017;9:981.
39. Tanaka S, Iimuro Y, Hirano T, Hai S, Suzumura K, Nakamura I, Kondo Y, Fujimoto J. Safety of hepatic resection for hepatocellular carcinoma in obese patients with cirrhosis. Surg Today. 2013;43(11):1290–7. https://doi.org/10.1007/s00595-013-0706-2.
40. Grąt M, Wronka KM, Krasnodębski M, Masior Ł, Lewandowski Z, Kosińska I, Grąt K, Stypułkowski J, Rejowski S, Wasilewicz M, Gałęcka M, Szachta P, Krawczyk M. Profile of gut microbiota associated with the presence of hepatocellular cancer in patients with liver cirrhosis. Transplant Proc. 2016;48(5):1687–91. https://doi.org/10.1016/j.transproceed.2016.01.077.
41. Amodio P, Bemeur C, Butterworth R, Cordoba J, Kato A, Montagnese S, Uribe M, Vilstrup H, Morgan MY. The nutritional management of hepatic encephalopathy in patients with cirrhosis: International Society for Hepatic Encephalopathy and Nitrogen Metabolism Consensus. Hepatology. 2013;58(1):325–36. https://doi.org/10.1002/hep.26370.

Gennaro Martines and Sebastio Perrino

9.1 Introduction

Obesity is a disease with a significant morbidity and mortality. About 3.4 million deaths related to obesity have been estimated worldwide in 2010, and its prevalence rose by 27.5% among adults between 1980 and 2013. Because of the far superior results of the surgical treatment compared to the medical therapy for obesity, the American Society for Metabolic and Bariatric Surgery has issued a grade A recommendation for bariatric surgery in patients with a body mass index (BMI) ≥ 40 kg/m^2 or for those with BMI ≥ 35 kg/m^2 and comorbidity unresponsive to previous medical treatment.

Bariatric surgery is the only medical branch where surgery can potentially solve multiple comorbidities (diabetes, hypertension, high cholesterolaemia, sleep apnoea, chronic cephalalgy, venostasis, urinary incontinence, liver diseases and arthritis) and is the only proven method that translates into a lasting weight loss. This well-established surgical approach, combined with the failure of the diet, the strong improvement of the quality of life and the fast postsurgical recovery with microinvasive techniques, determined an increase of the number of bariatric procedures performed yearly in the last 10 years [1].

Some historical documents report that the first bariatric surgical procedure was performed in the tenth century [2, 3]. D. Sancho, king of Leon, was described as obese to the point that he could not walk, ride a horse or use a sword. This led him to lose his throne. Escorted by his grandmother, he went to Cordoba, where he had his lips sutured by a famous Jewish doctor, Hasdai Ibn Shaprut. This way, the king

G. Martines (✉)
Surgical Unit, "M. Rubino"—Azienda Ospedaliero Universitaria Policlinico, Bari, Italy

S. Perrino
Endocrinology Unit, Department of Emergency and Organ Transplantation,
University Aldo Moro of Bari, Bari, Italy

© Springer Nature Switzerland AG 2019
D. F. Altomare, M. T. Rotelli (eds.), *Nutritional Support after Gastrointestinal Surgery*, https://doi.org/10.1007/978-3-030-16554-3_9

was only able to feed by a straw, on a liquid diet made of herbs. This procedure allowed him to lose half his body weight and gain his throne back.

The first metabolic surgery procedure—the jejunoileal bypass—was reported by Kremen in 1954 and consisted in an anastomosis between the proximal jejunum and the distal ileum in order to divert the food bolus through the small intestine. This procedure, however, was abandoned because of the occurrence of severe diarrhoea, malabsorption and dehydration.

As an alternative, a jejunocolic bypass was proposed by Henry Buchwald, with excellent results in lowering the lipid levels, even in the long term.

In 1966, Dr. Mason, after evaluating the loss of weight in patients undergoing subtotal gastrectomy for cancer, proposed the first real bariatric surgery procedure: the gastric bypass. It consisted in a horizontal gastric resection associated with an ileal loop anastomosis. Compared to the jejunoileal bypass, this technique had less complications.

In the 1990s bariatric surgery was consecrated as the main approach to treat obesity, now considered as an epidemic disease. Furthermore, thanks to the introduction of the laparoscopic techniques, the postoperative recovery became quicker and safer.

The procedures available today can be distinguished in "restrictive surgeries" (gastric banding, vertical gastroplasty, sleeve gastrectomy) and "malabsorptive surgeries" (biliopancreatic diversion according to Scopinaro and then modified by Gagner with the association of a vertical gastroplasty to a duodenal switch). The evolution continued during the twenty-first century, with the spread of the sleeve gastrectomy and of the mini-gastric bypass described by Rutledge.

The 2014 IFSO survey [4], published in 2017, reported the diffusion of sleeve gastrectomy, representing today about 45.9% of the bariatric surgery procedures, followed by the gastric bypass (39.6%), by the gastric banding (7.4%), by the mini-gastric bypass (1.8%) and finally by the biliopancreatic diversion (1.1%).

These data are confirmed by the survey made by the Italian Society of Bariatric Surgery and Metabolic Diseases (SICOB) in 2017 [5].

The endoscopic procedures developed in the last few years became very popular, both as bridge-to-surgery procedures (insertion of an intragastric balloon) and as one-step procedures, such as the endoscopic gastric plication (also called endoscopic sleeve), but still have contradictory outcomes [4].

No procedure can be considered as a gold standard, because the surgical choice has to be weighed against the characteristics of the patient, the surgeon's experience, the organizational level of the structure and the evaluation of the risk-benefit balance in the short and in the long term. In particular, a multidisciplinary team must be available and must be involved in a close follow-up, particularly for the malabsorptive procedures that cause an increased risk of micro and macro nutritional deficit in the short and in the long term [6].

9.2 The Physiology of Body Weight Regulation

The alimentary behaviour [7, 8] is determined by the homeostatic and reward centres of the brain, which integrate continuously the peripheral signals related to the energy deposits and nutrients availability (Fig. 9.1).

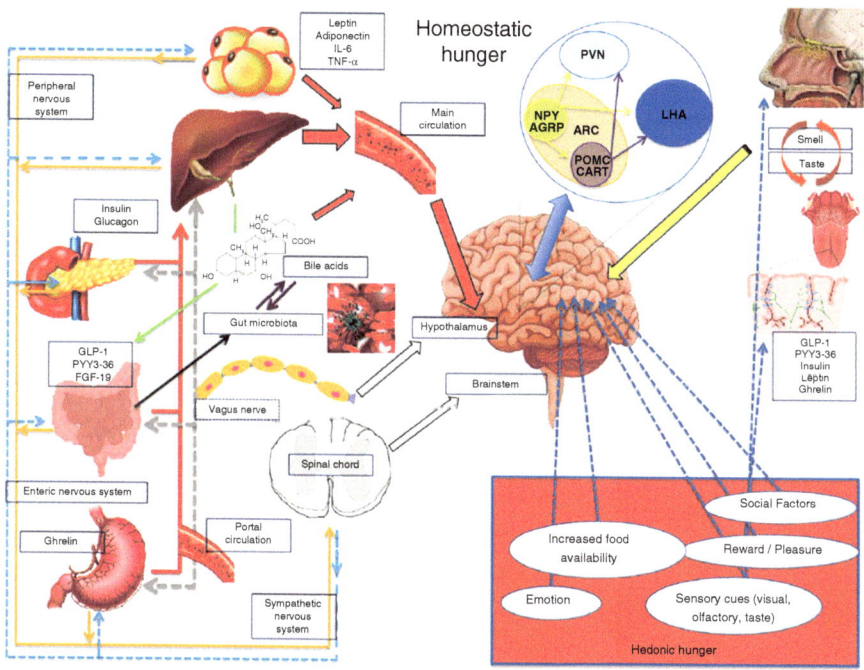

Fig. 9.1 Physiology of body weight regulation

Obesity occurs when the energy intake chronically exceeds the energy expenditure, which in turn can be due to an alteration of the homeostatic or hedonic system, or both [9].

The peripheral energetic signals are distinguished in long-term signals, such as leptin and insulin, providing information on the energy reservoir, and short-term signals, such as the availability messages related to the nutrients and the meals [8].

Intestinal hormones are produced by enteroendocrine gastrointestinal cells in response to the ingestion of nutrients and act as regulators of the energetic balance and glucidic homeostasis. The hormones peptide YY3-36 (PYY) and glucagon-like peptide-1 (GLP-1) are secreted by enteroendocrine L cells, located along the intestinal tract in response to the ingestion of nutrients. Both have an effect on the suppression of appetite by modulating neural activity within homeostatic and reward brain regions [10].

Furthermore, both PYY and GLP-1 influence the glycaemic regulation [11]. GLP-1 is one of the key mediators of the incretin effect (the increase of the secretion of insulin after the oral administration of glucose, as opposed to the endovenous administration) [12].

On the contrary, the ghrelin produced mainly by the P/D1 cells in the oxyntic glands of the fundus stimulates the appetite and the energetic provision [13]. Ghrelin levels increase quickly after the ingestion of food and decrease proportionally to the food intake in the postprandial period. Ghrelin acts also on the homeostatic and reward centres, and the increase of ghrelin can improve the hedonic responses to food [14]. Bile acids (BAs) are produced in the liver, stored in the gallbladder and released in the duodenum after the ingestion of food. They facilitate the formation of micelles and the absorption of the fats introduced with the diet and of the fat-soluble vitamins. They seem to have a role in the energetic and glucidic homeostasis [15].

BAs activate the secretion of GLP-1 by activating the G protein-coupled receptors (TGR5) on L cells, and fasting total circulating BA levels are positively correlated with postprandial GLP-1 levels [16]. BAs also act on farnesoid X receptor (FXR) [17] present in pancreatic β cells, increasing insulin release. BA activation of intestinal FXR cells stimulates the secretion of fibroblast growth factor-19 (FGF-19), a protein that contributes to improved peripheral glucose disposal and lipid homeostasis, leading to reduced weight and increased metabolic rate. In animal studies [18], BA supplementation has been shown to reduce weight gain, and postprandial BA levels are inversely related with body fat mass. Thus, the physiological effects of BA likely extend beyond the gut and pancreas with TGR5 receptors also located on skeletal muscles [19].

The human intestine hosts trillions of microorganisms. The gut microbiota can influence energy absorption by the alteration of the permeability of the intestinal mucosa, with energy consumption by the activation of the intracellular thyroid hormone through the FXR signal and the immunological systems of their human hosts [20]. Diet, exposure to antibiotics and other environmental factors can in turn influence the diversity of the microbiota and its function.

Taste and smell signals can affect the energy intake by influencing food selection. There is a close interaction between energetic homeostatic signals, taste and sense of smell. Insulin, leptin, GLP-1, PYY and ghrelin have been found in saliva, and their cognitive receptors have been identified on taste buds and olfactory neurons [21].

Sensory stimuli can cancel the satiety signals, leading to excess of energy. This brings to the deregulation of the homeostatic mechanisms that normally control the body weight by predisposing individuals to gaining more weight [14].

9.3 Physiopathological Alteration of Energetic Balance in the Obese Patient

Obesity is the result of a chronically positive energetic balance [22]. Once the state of obesity is fully established, many physiopathological changes occur, including resistance to leptin and insulin and a reduction pf PYY and GLP-1 in plasma as a response to the ingestion of nutrients. There is also a reduction in the postprandial suppression of ghrelin in the blood (Fig. 9.2).

Obesity has also been proven to mitigate the increase of the postprandial BA levels in the blood flow [23].

A dysbiotic relationship between host and microbiota has been suggested as a contribution to the development of obesity, with profound differences between the composition of the microbiota between obese and slim individuals [24].

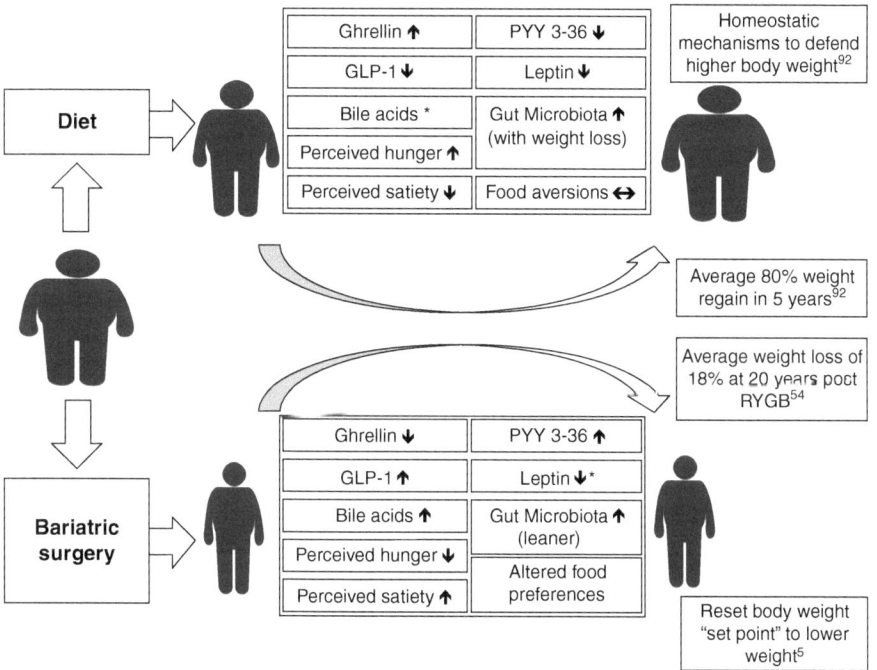

Fig. 9.2 Physiopathological alteration of the obese

9.4 Classification of the Surgical Procedures

9.4.1 Mechanical Restrictive Surgery

9.4.1.1 Laparoscopic Adjustable Gastric Band (LAGB)

This procedure was introduced in Europe by Kuzmak in the 1980s and popularized in the 1990s with the development of laparoscopic surgery [2].

It consists (Fig. 9.3) in the placement of a round silicone prosthesis around the upper portion of the stomach (about 4 cm from the cardias), adjustable by a tube connected to a port placed subcutaneously in the abdomen, allowing to obtain a small gastric pouch (about 30 ml) which will collect the food coming from the oesophagus.

Currently the most popular technique provides the passage of the prosthesis through the *pars flaccida* of the small omentum and has basically replaced the old perigastric technique, causing less complications.

The weight loss by LAGB is based on the reduced total energy intake, because of the significantly modified alimentary behaviour. Furthermore, these patients switch to a diet richer in softer food, with a reduction in the consumption of solid food.

Fig. 9.3 Adjustable
gastric band

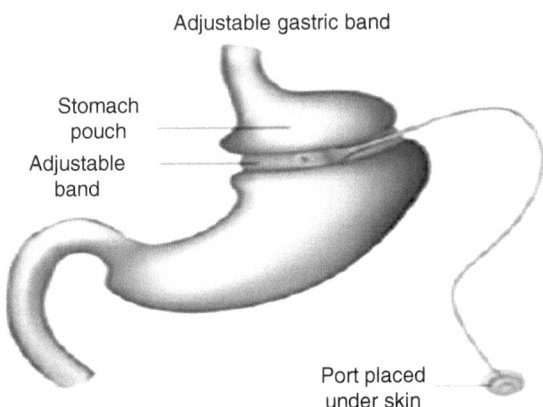

Adjustable gastric band

Stomach
pouch

Adjustable
band

Port placed
under skin

It has been shown that, with LAGB, it is impossible to eat food with a coarser texture, because it will not pass beyond the LAGB. Even the most stable patients, who obtained a positive result in terms of weight loss, sometimes regurgitate the food if the bolus is inappropriate.

Although these mechanical factors are imposed by LAGB, independently from the fact that they are necessary for the behavioural adaptations, the following loss of weight remains unexplained. Based on the results of other therapies, it seems unlikely that the purely mechanical factors can be able to bypass the control processes that rigidly protect the body weight. Therefore, other mechanisms are potentially responsible for the observed results.

LAGB constantly determines an intraluminal resistance just below the gastro-esophageal junction. This results in a delayed transit of semisolid food in the stomach, passing through the LAGB by repeated oesophageal peristalses. This strongly suggests that the preservation of food in a proximal restricted pouch is not the way LAGB works.

A better understanding of the main basic physiological intraluminal processes associated with LAGB is now available, particularly in terms of intraluminal transit, flow and dynamic, as well as basic correlations with sensations and results.

There has been a substantial interest for the gastrointestinal hormones involved in the regulation of food intake, appetite and energetic balance. Ghrelin levels seem to increase after LAGB, and this seems related to the weight loss induced by the diet. This may be an example of the ability of LAGB to maintain the suppression of appetite in the long term, overcoming the compensatory hormonal changes.

Each of the gastrointestinal hormones that influence satiety and food intake also has multiple physiological functions. Therefore, it remains critical to establish a true physiological role for peripheral satiety and hunger signals.

The future task is to link and expand these understandings to identify specifically the mechanism of weight loss [25].

9.4.2   Restrictive and Functional Surgery (Anorectic)

9.4.2.1 Sleeve Gastrectomy

This procedure consists in the vertical resection of about 4/5 of the stomach (with the use of linear laparoscopic staplers, the cartridges of which will be adapted to the thickness of the gastric portion to resect), by completely removing the fundus, keeping a distance of at least 1 cm from the His angle (Fig. 9.4). The size of the tubule depends on a calibrating bougie, which guides in the resection of the stomach [2].

The creation of the gastric tubule determines a lower food intake, but it seems to cause also important hormonal changes (such as decrease of the ghrelin, increase of GLP-1 and PYY) that tend to reduce the appetite of the obese patients.

The factors that cause the loss of weight after bariatric surgery are still uncertain, but it seems clear that, besides malabsorption and gastric restriction, other mechanics are involved.

In sleeve gastrectomy, besides the restriction of the gastric volume, there is also a decrease of the ghrelin (related to the complete removal of the fundus, where ghrelin-producing cells are located) with an increase of the hormonal levels of PYY and GLP-1, which however remain constant even after many years after the surgery [26].

9.4.2.2 Gastric Bypass

The Roux-en-Y gastric bypass consists in the creation of a very small gastric pouch, about 20–30 ml big, connected to the oesophagus, completely separated from the stomach remnant (which remains not accessible) and anastomosed with the small intestine.

A second anastomosis is performed between the food loop and the biliopancreatic loop. The length of these limbs is not standardized (Fig. 9.5).

Different versions of the procedure exist: bypass with fundectomy or banded bypass. If the surgeon wishes to keep an access to the papilla of Vater, he has to resort to the bypass on vertical gastroplasty [2].

Fig. 9.4 Sleeve gastrectomy

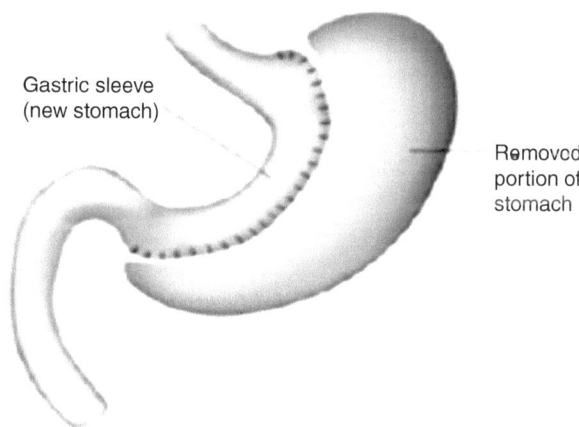

Gastric sleeve
(new stomach)

Removed
portion of
stomach

Fig. 9.5 Gastric bypass

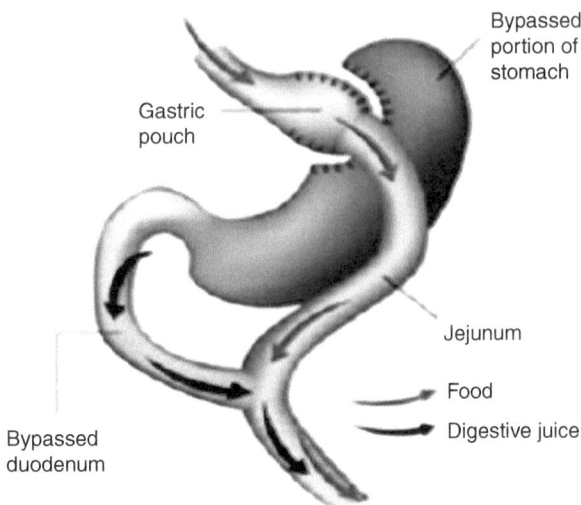

After the gastric bypass, an increased production of the GLP-1 incretin has been reported.

GLP-1 is metabolized immediately after its release by the dipeptidyl peptidase-4 enzyme (DPP-4), which cleaves the N-terminal dipeptide, which in turn releases a metabolite (GLP-1 9-36 amide) devoid of insulinotropic activity. This is biologically relevant, since immediately after its release from the L cell, the hormone interacts with sensorial vagal afferents and components of the enteric system before being degraded. Therefore, the metabolite has already performed its activity before being decomposed. Probably this mechanism explains the effect of GLP-1 on the secretion of insulin and on appetite after food intake. It has been reported that only 10–15% of the secreted GLP-1 reaches the peripheral target tissue in its nonmetabolized form. After the bypass procedure, the concentration of the nonmetabolized hormone is up to ten times higher. Therefore, it is possible that the endocrine route is more important after RYGB than it is in unoperated individuals.

It's also clear that the local concentrations of GLP-1 in the intestine, in the portal circulation and in the liver are much higher than normal and can lead to a more extensive activation of the sensorial neuronal paths.

The secretion of the glucose-dependant insulinotropic polypeptide (GIP) can slightly increase or decrease, and it does not show coherent patterns.

Presumably, whether GIP secretion is altered or not may depend on the length of the limbs of the Roux-en-Y reconstruction: the shorter the bypass of the proximal jejunum, the greater the GIP response.

The secretion of ghrelin, the hunger hormone, generally decreases immediately after RYGB, and its decrease could be related to the loss of appetite. However, after some months, the secretion of ghrelin goes back to normal and does not show further alterations in the long term. Therefore, ghrelin does not seem to play an essential role in the progressive loss of weight observed in the first 1–2 years, with the exception of the first 2 months. The concentration of cholecystokinin (CCK) shows a significant increase after the surgery. This is interesting in the light of its well-documented role of regulator of food intake. Furthermore, it is possible to observe an exaggerated secretion of the PYY 3-36 peptide, an active form of the PYY hormone in relation to food intake, that, as GLP-1, is produced by the L cells. It is evident that PYY, just like GLP-1, plays an important role in food intake inhibition. Gastrin responses are clearly decreased, probably consistently with the stomach bypass and suggesting that the luminal H+ concentration is high, whereas plasma somatostatin (including both the 1–14 and the 1–28 forms) does not show any meaningful changes. This is actually surprising since some of the somatostatin cells in the intestinal mucosa are clearly of the open type and would be expected to respond to luminal stimulation [26].

Dumping syndrome has been reported after bariatric surgery, mainly in patients who underwent a RYGB. The physiopathological mechanisms involved in the dumping syndrome remain unclear (Fig. 9.6).

Early dumping syndrome is the most frequent, and it can occur alone or together with late symptoms. The late isolated dumping (hypoglycaemia being its only symptom) affects up to 25% of the patients.

The symptoms suggesting an early and late dumping syndrome can be serious and can persist in some patients for many years after surgery.

Very few studies have been conducted recently with the aim to clarify the mechanisms involved in the early and late dumping syndrome. Most of our current knowledge is therefore based on older literature.

Several concurrent phenomena contribute to the development of early symptoms of dumping. Gastric surgery reduces the gastric volume or removes the barrier function of the pylorus, with the consequent fast delivery of an amount of solid not-digested food to the small intestine.

The sudden passage of hyperosmolar contents from the stomach presumably causes a shift of fluid from the intravascular compartment (i.e. plasma) to the intestinal lumen, with consequent plasma volume contraction, tachycardia and rarely syncope. The fluid shift in the small bowel can also cause distention and contribute to cramp-like contractions, swelling and diarrhoea. It is unknown whether this fluid shift contributes to the physiopathology of the dumping syndrome or is mainly a consequence of this process. In favour of the latter interpretation, intravenous fluid replacement is not effective in the prevention of early dumping syndrome. Another important mechanism of the physiopathology of early dumping syndrome (as well as of late dumping syndrome, as described later on) involves the increase of the release of several gastrointestinal hormones, including vasoactive agents (such as neurotensin and vasoactive intestinal peptide [VIP]), incretins (i.e. gastric inhibitory polypeptide [GIP] and GLP-1) and glucose modulators (i.e. insulin and glucagon) [27].

The increase in the release of these hormones can induce gastrointestinal motility and inhibit secretion, as well as elicit hemodynamic effects. For instance, neurotensin and vasoactive intestinal polypeptide induce a splanchnic vasodilation causing hypotension and systemic hemoconcentration. In contrast to the multiple pathophysiologic factors involved in early dumping syndrome, the physiopathology

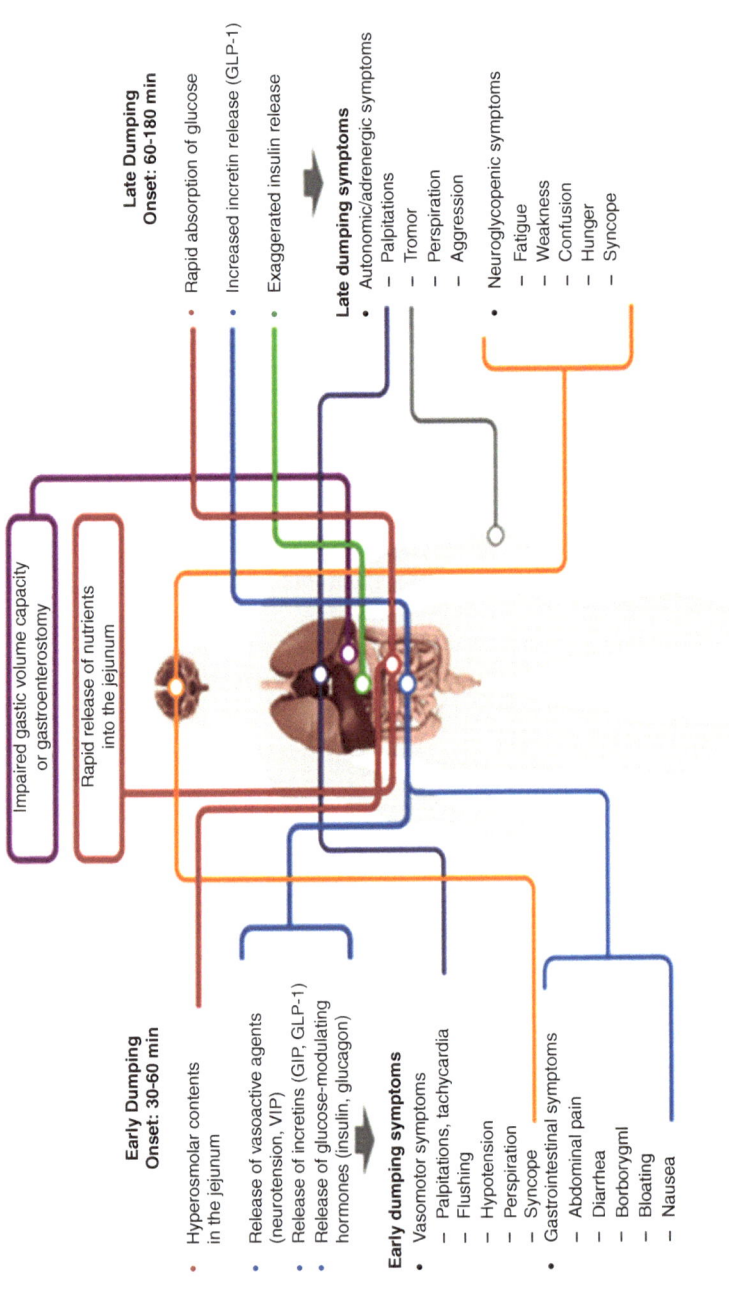

Fig. 9.6 Physiopathological mechanisms of dumping syndrome

Early Dumping
Onset: 30–60 min

– Hyperosmolar contents in the jejunum
– Release of vasoactive agents (neurotension, VIP)
– Release of incretins (GIP, GLP-1)
– Release of glucose-modulating hormones (insulin, glucagon)

Early dumping symptoms
- Vasomotor symptoms
  – Palpitations, tachycardia
  – Flushing
  – Hypotension
  – Perspiration
  – Syncope
- Gastrointestinal symptoms
  – Abdominal pain
  – Diarrhea
  – Borborygml
  – Bloating
  – Nausea

Impaired gastic volume capacity or gastroenterostomy

Rapid release of nutrients into the jejunum

Late Dumping
Onset: 60–180 min

– Rapid absorption of glucose
– Increased incretin release (GLP-1)
– Exaggerated insulin release

Late dumping symptoms
- Autonomic/adrenergic symptoms
  – Palpitations
  – Tromor
  – Perspiration
  – Aggression
- Neuroglycopenic symptoms
  – Fatigue
  – Weakness
  – Confusion
  – Hunger
  – Syncope

of the late form is largely ascribable to the development of hyperinsulinemic or reactive hypoglycaemia.

The fast delivery of undigested carbohydrates to the small bowel determines high concentrations of glucose, inducing a hyperinsulinemic response, with consequent hypoglycaemia and related late dumping symptoms.

Enteral glucose administration induces a greater release of insulin compared to the intravenous administration, a process known as incretin effect.

Two hormones, GIP and GLP-1, play a fundamental role in the incretin effect. An increased GLP-1 response has been reported in patients after gastric surgery, and a positive correlation has been observed between the increase in GLP-1 levels and insulin release. Literature data also suggest that the GLP-1 analogues may actually stabilize the glucose levels in patients with postprandial hypoglycaemia after gastric bypass. Therefore, an exaggerated endogenous GLP-1 response seems to be the main cause of the hyperinsulinemic and hypoglycaemic effect that is characteristic of late dumping syndrome. However, the exact mechanism by which GLP-1 contributes to glucose homeostasis and late dumping syndrome remains to be fully cleared [27].

9.5   Malabsorptive Surgery

9.5.1   Biliopancreatic Diversion According to Scopinaro

This procedure consists in the creation of a gastric pouch (about 400 mL big) by resecting 2/3 of the distal stomach. The duodenum is resected at 2–3 cm from the pylorus and the small bowel is resected while kept at the maximum tension, 300 cm from the ileocecal valve. A termino-terminal gastro-ileal anastomosis and an ileo-ileal anastomosis are performed, in order to obtain an alimentary limb of 250 cm and a common limb of 50 cm (Fig. 9.7). A prophylactic cholecystectomy is performed.

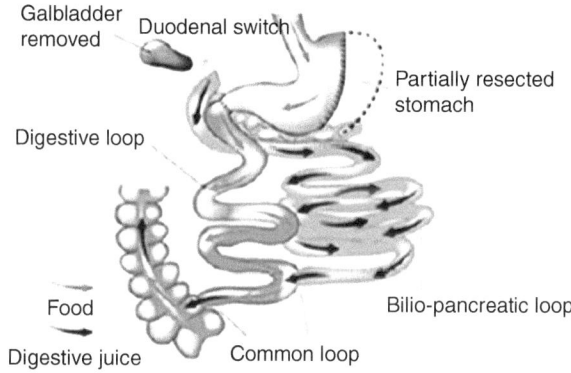

Fig. 9.7 Biliopancreatic diversion

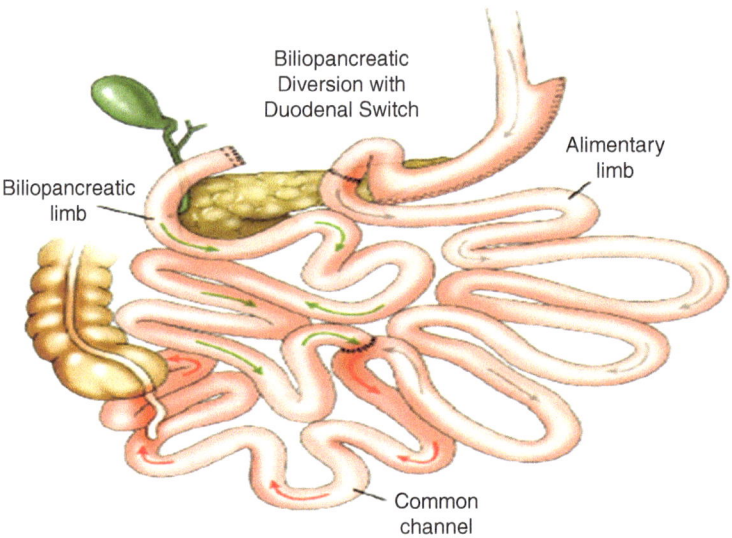

Fig. 9.8 Biliopancreatic diversion with duodenal switch

A variation of the biliopancreatic diversion is the duodenal switch (Fig. 9.8). Unlike the Scopinaro procedure, a vertical gastric pouch is created, like in sleeve gastrectomy.

This procedure consists in creating two separate pathways, one for food and the other for biliopancreatic juices, which join again near the colon.

The ingested food skips the biliopancreatic limb and is not digested by the biliary action and the pancreatic juices; therefore the fats are not digested nor absorbed.

Since pancreatic secretion is necessary to digest fats and starch, and digestion is necessary for their intestinal absorption, the result is a limitation of the absorption of these foods, which are the ones with the highest calorie content.

The whole process is regulated by the same enterohormonal mechanisms of the gastric bypass. This way, in all of the cases (100%), there is a resolution of type 2 diabetes.

Insulin resistance is quickly reversed, thanks to an extremely reduced daily lipid absorption with consequent intramyocellular fat storage reduction. Furthermore, there is a reduction of the serum cholesterol. This is caused by the reduction of the absorption of the bile salts, which forces the liver to increase bile acid synthesis using as base substance the cholesterol molecule itself, and by reduction of the absorption of the endogenous cholesterol, which is mostly contained in the bile (70% of the total cholesterol passing through the alimentary limb). Since cholesterol is absorbed together with fats, and the absorption of fats is strongly reduced, also the absorption of cholesterol is reduced accordingly.

Sugars and proteins are surely digested and absorbed less than usual, even if the reduction is not as big as with fats. This happens because a partial digestion with a partial absorption occurs in the residual limb of the small bowel that has been surgically connected directly to the stomach, skipping the duodenum. Patients who undergo this procedure absorb only a small part of the ingested sugar.

Obviously, these patients ingest bigger quantities of sugar in comparison to normal individuals, because of the scarce absorption. This amount quickly reaches the distal limb of the ileum, where the L cells produce GLP-1 incretin, which inhibits the glucagon-producing alpha cells and determines an increased insulin release and production, with trophic effects on islet beta cells.

Even though the results obtained by the biliopancreatic diversion in terms of loss of weight and resolution of type 2 diabetes are clear, this operation leads to important deterioration of the quality of life of these patients because of diarrhoea, serious anaemia, osteoporosis, etc. [28].

9.5.2   Mini-Gastric Bypass (One Anastomosis Gastric Bypass)

This operation consists in the creation of a small gastric pouch (60 ml) which is separated from the stomach remnant and connected to the small bowel by termino-lateral anastomosis at a distance of about 200 cm from the duodenum (Fig. 9.9).

This is a still non-standardized procedure, the results of which are still under evaluation [29].

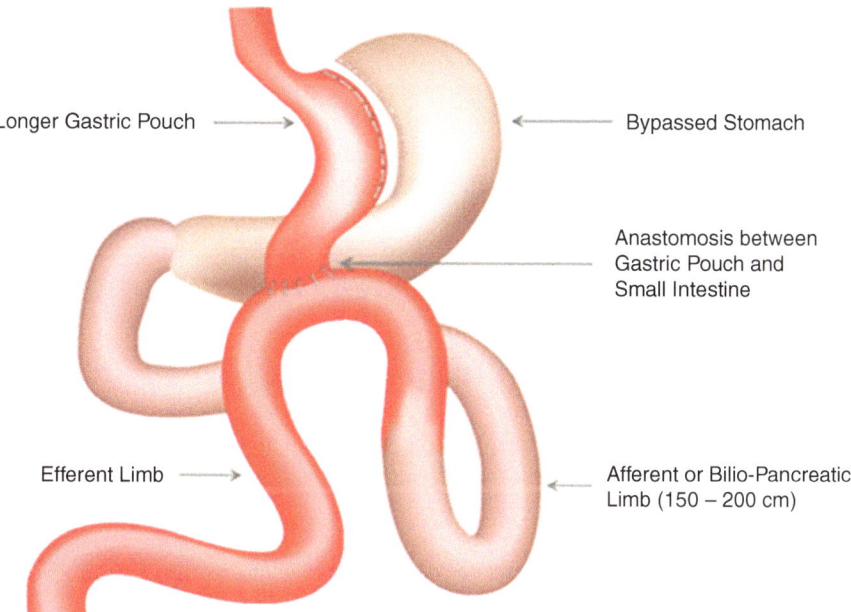

Fig. 9.9 One anastomosis gastric bypass

9.6   Endoscopic Procedures

9.6.1   Intragastric Balloon

In 1985, the Garren-Edwards gastric bubble (GEGB) was the first intragastric balloon approved for obesity treatment and was introduced in the US market.

The most frequently used balloon is the Orbera Intragastric Balloon (Apollo Endosurgery, Austin, TX, United States), which was known previously as the BioEnterics Intragastric Balloon (BIB). It is an elastic silicone balloon containing saline (450–700 mL). The positioning assembly, which comprises a balloon-filling tube and a catheter with the deflated balloon, is blindly advanced to the gastro-oesophageal junction. An endoscopic device is inserted to ensure the precise deployment of the intragastric balloon, which is then filled with methylene-mixed saline under direct observation via the catheter. If an unexpected balloon rupture occurs, the methylene blue turns the urine green (Fig. 9.10). The Orbera balloon is usually implanted for 6 months, removed endoscopically by needle aspiration of the intragastric fluid, and retrieved with a snare or grasper. The FDA approved the use of the Orbera balloon on August 6, 2015. It is expected that the Orbera balloon could provide a valuable and less invasive therapeutic approach to bariatric treatment.

9.6.1.1 Physiopathological Factors

The purpose of the treatment by intragastric balloon is stimulating weight loss and helping to improve the associated comorbidities, with an appropriate safety. The restriction of the gastric capacity is an essential factor in bariatric surgery. It induces early satiety and strengthens the stimulation of the gastric mechanics and chemistry by interacting with various exogenous gastric factors. It affects hunger control and gastric emptying by the alteration of intestinal hormones and peptides.

The intragastric balloon imitates the surgical procedure by reducing the effective gastric volume. Furthermore, the treatment by gastric balloon can influence the changes in weight through the interaction of neuronal gastric factors. These factors

Fig. 9.10 Intragastric balloon

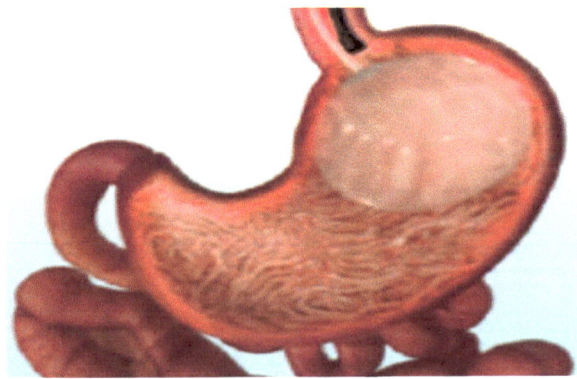

include several intestinal peptides and hormones, such as ghrelin, leptin, cholecys-tokinin, YY peptide, pancreatic polypeptide and GLP-1.

A study with 40 obese patients who underwent balloon placement indicated no effect on ghrelin levels when patients were fasting or meal-suppressed. In another study, 17 patients with non-morbid obesity underwent balloon placement, and fast-ing plasma ghrelin concentrations significantly decreased (3.2–1.9 ng/mL; $p = 0.021$) as a result. Martinez-Brocca et al. reported that fasting and meal-suppressed plasma ghrelin levels did not differ significantly between groups in mor-bidly obese patients. Konopko-Zubrzycka et al. reported that body weight reduction after balloon treatment is related to a transient elevation in plasma ghrelin levels and a decrease in plasma leptin levels.

Another study evaluated fasting and postprandial cholecystokinin and pancreatic polypeptide secretion after 13 weeks of balloon treatment in obese patients. Baseline and meal-stimulated cholecystokinin levels were decreased (Table 9.1).

These results of course are absolutely temporary if they are not followed by sur-gical treatment [30].

9.7   Nutritional Support After Bariatric Surgery

Bariatric surgery provides substantial and sustained effects on weight loss and ame-liorates obesity-attributable comorbidities in the majority of bariatric patients, although risks of complication, reoperation and death exist [31]. While a patient's surgeon monitors closely for postoperative surgical complications, the primary care provider or endocrinologist is often the provider to identify and manage postopera-tive medical and nutritional complications. This chapter reviews these potential nutritional complications with attention to screening and therapeutic approach.

9.7.1   Early and Late Nutritional Management

Most bariatric procedures include the reduction of the volume of the stomach and/or the creation of a small gastric pouch. Therefore, the ingestion of solid foods in the first days after surgery is impossible, and a gradual change of food consistency in the first postoperative weeks is preferred in order to avoid or minimize regurgitation and vomiting, which can threaten the integrity and safety of the recent surgical pro-cedure and result in severe vitamin B1 (thiamine) deficiency [32]. After bariatric surgery, a low-sugar clear liquid meal programme is usually initiated within 24 h, and patients are then informed to gradually and progressively change the food con-sistency, moving from clear liquids to soft or creamy foods and then to solid chew-able items over a period of 2–4 weeks [32, 33]. The aim of dietary counselling should be the fitting of patients' eating behaviour to the surgical procedure and the general qualitative aspects of a healthy nutrient-dense diet. In detail, patients with gastric restriction should be counselled to eat three small meals during the day and chew small bites of food thoroughly before swallowing, without drinking beverages

Table 9.1 Intragastric balloon: physiopathological factors

Refs.	Balloon	Weight loss at 6 months	Ghrelin T0	Ghrelin (after 3 months)	Ghrelin (after6 months)	Ghrelin (after 12 months)	Leptin T0	Leptin (after 3 months)	Leptin (after 6 months)	Leptin (after 12 months)
Mathus VIiegen et al.	Orbera	17.4 ± 7.8 kg	722.31 ± 151.5 pg/mL	791.51 ± 239.0 pg/mL	743.7 ± 115.2 pg/mL	N/A	N/A	N/A	N/A	N/A
Fuller et al.	N/A	14.2%	414.1 pmol/L	448 pmol/L	452.4 pmol/L	379.4 pmol/L	23.4 ng/mL	18.5 ng/mL	11.7 ng/mL	19.7 ng/mL
Bużga et al.	MedSil®	184 ± 18.2 kg	240.5 ± 101.5 µg/L	378.1 ± 155.8 µg/L	335.8 ± 149.2 µg/L	N/A	30.4 ± 17.2 µg/L	18.2 ± 15.8 µg/L	14.9 ± 15.5 µg/L	N/A
Nikolic et al.	Orbera	N/A	958.3 pg/mL	1346.2 pg/mL	1050.1 pg/mL	922.6 pg/mL	25.1 ng/mL	14.3 ng/mL	10.5 ng/mL	17.5 ng/mL
Konopko-Zubrzyea et al.	Orbera	17.1 ± 8.0 kg	621.91 ± 182.4 pg/mL	903.9 ± 237 pg/mL	N/A	N/A	61.3 ± 36.7 ng/mL	39.9 ± 17.5 ng/mL	N/A	N/A
Martinez-Brocca et al.	Orbera	12.7 ± 5.6 kg (4 months)	934.4 ± 199.2 pg/mL	947.1 ± 195.1 pg/mL	N/A	N/A	31.9 ± 164 ng/mL	22.4 ± 15.1 ng/mL	N/A	N/A
Mion et al.	Orbera	8.7 kg	3.2 ± 0.4 ng/mL	N/A	1.9 ± 0.1 ng/mL	N/A	27.8 ± 3.7 ng/mL	N/A	18.7 ± 27 ng/mL	N/A

Data from patients with body mass indexes <40 kg/m^2. Data are presented as mean ± SD or median. N/A: not available

Ref. [30]

at the same time (more than 30 min apart) [32, 33]. Regular physical activity is considered a critical factor for weight maintenance and should therefore be encouraged after bariatric surgery. Patients should be advised to carry out moderate aerobic physical activity, i.e. a minimum of 150 min/week (with a goal of 300 min/week), as well as strength training 2–3 times/week [32, 33].

9.7.1.1 Protein Supplementation

Protein intake is generally reduced following bariatric surgery, and then adequate protein intake is prominent to counteract the loss of lean body mass in any situation when a rapid weight loss occurs [34]. Dietary counselling should address the problem of protein intake, particularly in the first months after surgery. Current guidelines suggest a minimal protein intake of 60 g/day and up to 1.5 g/kg ideal body weight per day, but higher amounts of protein intake (up to 2.1 g/kg ideal body weight per day) may be required in individual cases [32–36]. The use of liquid protein supplements (30 g/day) can facilitate adequate protein intake in the first period after surgery. Bariatric procedures involving a certain degree of malabsorption can cause protein malnutrition. The incidence of protein malnutrition depends on the degree of the malabsorption as well as on the dietary habits and the protein requirements of the patients. A 13% incidence has been reported after distal Roux-en Y gastric bypass (RYGB) with a Roux limb ≥150 cm, whereas an incidence of protein malnutrition ranging from 3 to 18% has been reported after biliopancreatic diversion [33]. Prevention of protein malnutrition involves regular rating of protein intake, fostering the ingestion of protein-rich foods (>60 g/day) divided into several meals and the use of modular protein supplements [32, 33]. In case of severe nonresponsive protein malnutrition, parenteral nutrition is recommended, and surgical revision with lengthening of the common channel to decrease malabsorption should be considered if a patient remains dependent on parenteral nutrition or has recurrent episodes of protein depletion [32–34].

9.7.1.2 Micronutrient Deficiencies

Given the dietary changes, rerouting of nutrient flow and gut anatomy/physiology alterations that occur after bariatric surgery, patients who undergo these procedures are at risk for micronutrient deficiencies. Some of these deficiencies can result in severe consequences, such as neuropathy, heart failure and encephalopathy. Therefore, it is extremely important that patients understand the need for lifelong supplementation. Patients who have malabsorptive procedures, such as Roux-en-Y gastric bypass (RYGB) or biliopancreatic diversion with duodenal switch (BPD/DS), are at highest risk for micronutrient deficiencies and require a more extensive preoperative nutritional evaluation and postoperative monitoring and supplementation. But even with restrictive procedures, decreased oral intake and poor tolerance to certain food groups may also increase the risk for micronutrient deficiencies [1, 32].

Tables 9.2, 9.3, and 9.4 represent recommendations that have been adapted from the American Society for Metabolic and Bariatric Surgery (ASMBS) Integrated Health Nutrition Guidelines [35]; Clinical Practice Guidelines from the combined

Table 9.2 Recommended postoperative supplementation of vitamins and minerals

Micronutrient	Supplementation
Within a multivitamin with minerals product	
Thiamine	12 mg/day
Vitamin B$_{12}$ (cobalamin)	Oral or sublingual: 350–500 µg/dayIntranasal: 1000 µg/week Intramuscular: 1000 µg/month
Folate (folic acid)	400–800 µg/dayWomen of childbearing age: 800–1000 µg/day
Iron	18 mg/day elemental ironRYGB, SG, BPD/DS or menstruating women: 45–60 mg/dayTake separately from calcium supplements
Vitamin D	D$_3$ 3000 IU/day
Vitamin A	LAGB: vitamin A 5000 IU/dayRYGB or SG: vitamin A 5000–10,000 IU/dayBPD/DS: vitamin A 10,000 IU/day
Vitamin E	15 mg/day
Vitamin K	LAGB, SG or RYGB: 90–120 µg/day BPD/DS: 300 µg/day
Zinc	SG or LAGB: 8–11 mg/dayRYGB: 8–22 mg/dayBPD/DS: 16–22 mg/day
Copper	SG or LAGB: 1 mg/dayRYGB or BPD/DS: 2 mg/day
As separate supplementation	
Calcium	LAGB, SG, RYGB: calcium 1200–1500 mg/day (diet + supplements) BPD/DS: calcium 1800–2400 mg/day (diet + supplements) (as calcium citrate, in divided doses)

Table 9.3 Schedule for postoperative micronutrient monitoring

	6 months	12 months	18 months	24 months	Annually
Vitamin B$_{12}$	X	X	X	X	X
Folate	X	X	X	X	X
Iron, ferritin	X	X	X	X	X
25-hydroxyvitamin D	X	X	X	X	X
Calcium	X	X	X	X	X
Intact PTH	X	X	X	X	X
24-h urinary calcium	X	X		X	X
Thiamine	Optional	Optional	Optional	Optional	Optional
Vitamin A				Optional	Optional
Zinc	Optional	Optional		Optional	Optional
Copper		Optional			Optional

American Association of Clinical Endocrinologists (AACE), The Obesity Society (TOS), and ASMBS [36]; and the Endocrine Society Clinical Practice Guidelines [37]. These recommendations for adults reflect general guidelines, and patients with specific diseases may require further evaluation and closer monitoring. For example, in malabsorptive bariatric surgical procedures, the resulting nutritional anaemias beyond that an appropriate iron repletion might also require other micronutrient deficiencies in vitamin B12, folate, protein, copper, selenium and zinc, and these should be evaluated.

Pre-existing micronutrient deficiencies would be corrected prior to surgery in order to avoid clinically symptomatic or severe disease. For example, suboptimal

Table 9.4 Repletion recommendations for micronutrient deficiencies

Micronutrient	Repletion recommendation
Thiamine	Oral: 100 mg 2–3 times dailyIM: 250 mg daily for 3–5 days or 100–250 mg monthlyIV: 200 mg 2–3 times daily to 500 mg 1–2 times daily for 3–5 days, followed by 250 mg/day for 3–5 days Severe disease: administer thiamine prior to dextrose-containing solutions
Vitamin B$_{12}$(cobalamin)	Oral: 1000 µg/dayIM: 1000 µg/month to 1000–3000 µg/6–12 months
Folate (folic acid)	1000 µg/day orally
Iron	150–200 mg elemental iron/day, up to 300 mg 2–3 times daily Calcium may impair iron absorptionConsider co-administration of vitamin C to enhance absorption Consider IV iron infusions for severe/refractory iron deficiency
Vitamin D	D$_3$ 6000 IU/day or D$_2$ 50,000 IU 1–3 times per week, or more if needed to achieve and maintain 25-hydroxyvitamin D > 30 ng/mL
Calcium	Increase dose and titrate to normalize PTH ± 24-h urinary calcium level
Vitamin A	10,000–25,000 IU/day orally until clinical improvement (1–2 weeks) With corneal changes: 50,000–100,000 IU IM × 3 days, then 50,000 IU/day IM for 2 weeks
Vitamin E	Optimal therapeutic dose not clearly defined, consider 100–400 IU/day
Vitamin K	Acute malabsorption: 10 mg parentally Chronic malabsorption: 1–2 mg/day orally or 1–2 mg/week parentally
Zinc	There is insufficient evidence to make a dose-related recommendation
Copper	Mild-moderate deficiency: oral copper gluconate or sulphate 3–8 mg/day Severe deficiency: 2–4 mg/day of intravenous copper × 6 days

IM intramuscular, *IV* intravenous

levels of 25-hydroxyvitamin D are particularly common and may require supplementation prior to surgery.

It is universally accepted that post-bariatric supplementation (Table 9.2) is an important component of postoperative care. For example, vitamin B12 deficiency is common after RYGB without adequate supplementation, and oral doses of 350 µg/day have been shown to maintain normal plasma B12 levels. Other suggested micronutrient doses are either based on expert opinion or are similar to the recommended dietary allowance (RDA) [35–37].

Most micronutrients are provided in multivitamins, and chewable multivitamins are recommended postoperatively. Multivitamins for the general population can be used, provided that attention is paid to the product's micronutrient contents. The ASMBS recommends one general multivitamin tablet daily for patients who have had laparoscopic adjustable gastric banding (LAGB) or two general multivitamin tablets daily for those undergoing sleeve gastrectomy (SG), RYGB or BPD/DS. As an alternative to general multivitamins, bariatric surgery-specific, high-potency multivitamins are available and often contain the recommended doses of micronutrients in one tablet daily [35].

Multivitamins do not contain the recommended doses of calcium, as calcium can impede the absorption of other micronutrients. Therefore, separate calcium supplementation is usually required. Calcium citrate is the preferred form of supplemental calcium, as it is better absorbed than calcium carbonate in the state of impaired gastric acid production. A patient's dietary calcium intake should be considered when determining the dose of a calcium supplement, as the recommended intakes are generally total daily intakes (diet plus supplements) [35]. Iron absorption may be enhanced by co-administration of vitamin C (500–1000 mg) to create an acidic environment or when taken with meat. If inadequate absorption or intolerance occurs, parenteral iron replacement may be necessary.

A suggested schedule for postoperative biochemical monitoring is listed in Tables 9.3 and 9.4. Patients who develop micronutrient deficiencies may need more frequent monitoring.

Examinations should be performed after RYGB or BPD/DS. All of these could be suggested for patients submitted to restrictive surgery where frank deficiencies are less common.

Some surgeons perform additional early biochemical evaluation 3 months postoperatively, and the AACE/TOS/ASMBS Clinical Practice Guidelines suggest evaluation earlier than 6 months for some micronutrients [36].

Oral repletion is often sufficient for correcting micronutrient deficiencies, although parenteral therapy may be required in severe disease. After a repletion course, biochemical testing should be performed and a maintenance dose should be established. Micronutrient deficiencies may coexist; for example, malabsorptive procedures may result in deficiencies of the fat-soluble vitamins A, E and K [35–37].

9.7.1.3 Dumping Syndrome

Dumping syndrome was believed to be typical of gastric bypass (70–75% of patients in the first year after surgery) [36], but it has been described also after sleeve gastrectomy (40% of patients 6 months after surgery) [37]. Early dumping syndrome typically occurs within 1 h of eating and is characterized by both gastrointestinal (nausea, abdominal fullness, diarrhoea) and vasomotor symptoms (fainting, sleepiness, weakness, diaphoresis, palpitations and desire to lie down) [38]. Dumping syndrome symptoms can appear as early as 6 weeks after surgery and has been reported to affect up to 20% according to large survey studies and up to 40% in smaller prospective studies of individuals who have undergone both restrictive and malabsorptive procedures [39–42]. The pathophysiology of dumping syndrome is not completely understood but is thought to be due to both a rapid delivery of nutrients to the small intestine causing an osmotic shift of intravascular fluid to the intestinal lumen and an increased release of gastrointestinal hormones that disrupt motility and hemodynamic status [43]. There is debate in the literature on whether dumping syndrome is an adaptive consequence of bariatric surgery that helps restrict food intake and aids weight loss versus an adverse consequence that reduces quality of life and does not contribute to weight loss [39].

The diagnosis of dumping syndrome should be made after the exclusion of more serious entities such as intestinal fistulas, adhesions, ischaemia, herniation, obstipation and gallstone disease which may have shared clinical features [43]. There are validated questionnaires as well as provocation tests that have been used to confirm dumping syndrome in research settings. Oral glucose challenge with an increase in heart rate and haematocrit (indicating hemoconcentration) is one such approach [38].

The first-line treatment for dumping syndrome is to modify the diet so as to avoid foods that worsen symptoms (oftentimes calorie-dense foods with high fat/refined sugar content and low in fibre), eating small volume meals, not eating and drinking at the same time, eating slowly, chewing well and avoiding alcohol. Indeed, patients often implement these changes on their own and, over time, symptom severity improves or resolves in many (if not most) patients. In addition, lying down for 30 min after eating to slow gastric emptying and mitigate symptoms of hypovolaemia may be helpful if symptoms occur [43]. There are several small interventional studies and case reports that support the use of dietary supplements (e.g. pectin, guar gum) that increase food viscosity and reduced symptoms of dumping syndrome; however, low palatability and potential choking hazard and bowel obstruction are downsides to their use [43]. Somatostatin analogues have also been tested in small studies, although this class of drugs are expensive, involve subcutaneous or intramuscular injections and have gastrointestinal side effects [43]. Enteral tube feedings or bariatric surgery reversal have been reported to improve symptoms when all else fails [43].

9.8 Conclusions

The postoperative management of the bariatric surgery patient requires a special knowledge and skills of the clinicians in order to deliver appropriate and effective care to the post-bariatric patient. Thus, an interdisciplinary team including the surgeon, dietitian and endocrinologist and/or primary care provider should provide a follow-up programme as an integral part of the clinical pathway at centres delivering bariatric surgery. This follow-up programme should include the management of chronic metabolic conditions and the prevention and treatment of postoperative medical and nutritional complications to optimize the long-term benefits of bariatric surgery.

References

1. Mechanick JI, Youdim A, Jones DB, Garvey WT, Hurley DL, McMahon MM, Heinberg LJ, Kushner R, Adams TD, Shikora S, Dixon JB, Brethauer S, American Association of Clinical Endocrinologists; Obesity Society; American Society for Metabolic & Bariatric Surgery. Clinical practice guidelines for the perioperative nutritional, metabolic, and nonsurgical support of the bariatric surgery patient—2013 update: cosponsored by American Association of Clinical Endocrinologists, The Obesity Society, and American Society for Metabolic & Bariatric Surgery. Obesity (Silver Spring). 2013;21(Suppl 1):S1–S27.
2. Faria GR. A brief history of bariatric surgery. Porto Biomed J. 2017;2;90–2. https://doi.org/10.1016/j.pbj.2017.01.008.
3. Story of Obesity Surgery. https://asmbs.org/resources/story-of-obesity-surgery
4. Angrisani L, Santonicola A, Iovino P, Vitiello A, Zundel N, Buchwald H, Scopinaro N. Bariatric surgery and endoluminal procedures: IFSO Worldwide Survey 2014. Obes Surg. 2017;27(9):2279–89.

5. Italian Society of Bariatric Surgery and Metabolic Diseases (S.I.C.O.B.). Survey 2017. https://www.sicob.org/area_04_medici/00_indagine.aspx.
6. Italian Society of Bariatric Surgery and Metabolic Diseases (S.I.C.O.B.). Guidelines. Edition 2016. https://www.sicob.org/area04medici/40lineeguida.aspx
7. Gloy VL, Briel M, Bhatt DL, et al. Bariatric surgery versus non-surgical treatment for obesity: a systematic review and meta-analysis of randomised controlled trials. BMJ. 2013;347:f5934. https://doi.org/10.1136/bmj.f5934.
8. Miras AD, le Roux CW. Mechanisms underlying weight loss after bariatric surgery. Nat Rev Gastroenterol Hepatol. 2013;10(10):575–84.
9. Makaronidis JM, Batterham RL. Obesity, body weight regulation and the brain: insights from fMRI. Br J Radiol. 2018;91(1089):20170910. https://doi.org/10.1259/bjr.20170910.
10. Scott WR, Batterham RL. Roux-en-Y gastric bypass and laparoscopic sleeve gastrectomy: understanding weight loss and improvements in type 2 diabetes after bariatric surgery. Am J Phys Regul Integr Comp Phys. 2011;301(1):R15–27.
11. Batterham RL, Cowley MA, Small CJ, et al. Gut hormone PYY(3-36) physiologically inhibits food intake. Nature. 2002;418(6898):650–4.
12. Guida C, Stephen S, Guitton R, et al. The role of PYY in pancreatic islet physiology and surgical control of diabetes. Trends Endocrinol Metab. 2017;28(8):626–36.
13. Muller TD, Nogueiras R, Andermann ML, et al. Ghrelin. Mol Metab. 2015;4(6):437–60.
14. Berthoud HR. Metabolic and hedonic drives in the neural control of appetite: who is the boss? Curr Opin Neurobiol. 2011;21(6):888–96.
15. Seeley RJ, Chambers AP, Sandoval DA. The role of gut adaptation in the potent effects of multiple bariatric surgeries on obesity and diabetes. Cell Metab. 2015;21(3):369–78.
16. Thomas C, Gioiello A, Noriega L, et al. TGR5-mediated bile acid sensing controls glucose homeostasis. Cell Metab. 2009;10(3):167–77.
17. Dufer M, Horth K, Wagner R, et al. Bile acids acutely stimulate insulin secretion of mouse beta-cells via farnesoid X receptor activation and K(ATP) channel inhibition. Diabetes. 2012;61(6):1479–89.
18. Watanabe M, Houten SM, Mataki C, et al. Bile acids induce energy expenditure by promoting intracellular thyroid hormone activation. Nature. 2006;439(7075):484–9.
19. Suzuki T, Aoyama J, Hashimoto M, et al. Correlation between postprandial bile acids and body fat mass in healthy normal-weight subjects. Clin Biochem. 2014;47(12):1128–31.
20. Martinez KB, Pierre JF, Chang ED. The gut microbiota: the gateway to improved metabolism. Gastroenterol Clin N Am. 2016;45(4):601–14.
21. Cummings DE. Taste and the regulation of food intake: it's not just about flavor. Am J Clin Nutr. 2015;102(4):717–8.
22. Piaggi P, Vinales KL, Basolo A, et al. Energy expenditure in the etiology of human obesity: spendthrift and thrifty metabolic phenotypes and energy-sensing mechanisms. J Endocrinol Investig. 2018;41(1):83–9.
23. Glicksman C, Pournaras DJ, Wright M, et al. Postprandial plasma bile acid responses in normal weight and obese subjects. Ann Clin Biochem. 2010;47(Pt 5):482–4.
24. Aron-Wisnewsky J, Clement K. The effects of gastrointestinal surgery on gut microbiota: potential contribution to improved insulin sensitivity. Curr Atheroscler Rep. 2014;16(11):454.
25. Burton PR, Brown WA. The mechanism of weight loss with laparoscopic adjustable gastric banding: induction of satiety not restriction. Int J Obes. 2011;35:S26–30.
26. Pucci A, Batterham RL. Mechanisms underlying the weight loss effects of RYGB and SG: similar, yet different. J Endocrinol Investig. 2019;42(2):117–28. https://doi.org/10.1007/s40618-018-0892-2.
27. van Beek AP, Emous M, Laville M, Tack J. Dumping syndrome after esophageal, gastric or bariatric surgery: pathophysiology, diagnosis, and management. Obes Rev. 2017;18:68–85.
28. Scopinaro N. Thirty-five years of biliopancreatic diversion: notes on gastrointestinal physiology to complete the published information useful for a better understanding and clinical of the operation. Obes Surg. 2012;22(3):427–32.

29. Musella M, Apers J, Rheinwalt K, Ribeiro R, Manno E, Greco F, Čierny M, Milone M, Di Stefano C, Guler S, Van Lessen IM, Guerra A, Maglio MN, Bonfanti R, Novotna R, Coretti G, Piazza L. Efficacy of bariatric surgery in type 2 diabetes mellitus remission: the role of mini gastric bypass/one anastomosis gastric bypass and sleeve gastrectomy at 1 year of follow-up. A European Survey. Obes Surg. 2016;26(5):933–40. https://doi.org/10.1007/s11695-015-1865-6.

30. Kim SH, Chun HJ, Choi HS, Kim ES, Keum B, Jeen YT. Current status of intragastric balloon for obesity treatment. World J Gastroenterol. 2016;22(24):5495–504.

31. Chang S-H, Stoll CRT, Song J, Varela JE, Eagon CJ, Colditz GA. Bariatric surgery: an updated systematic review and meta-analysis, 2003–2012. JAMA Surg. 2014;149(3):275–87.

32. Busetto L, Dicker D, Azran C, Batterham RL, Farpour-Lambert N, Fried M, et al. Obesity management task force of the European Association for the study of obesity released "practical recommendations for the post-bariatric surgery medical management". Obes Surg. 2018;28(7):2117–21.

33. Mechanick JI, Kushner RF, Sugerman HJ, Gonzalez-Campoy JM, Collazo-Clavell ML, Spitz AF, et al. American Association of Clinical Endocrinologists, The Obesity Society, and American Society for Metabolic & Bariatric Surgery medical guidelines for clinical practice for the perioperative nutritional, metabolic, and nonsurgical support of the bariatric surgery patient. Obesity (Silver Spring). 2009;17(Suppl 1):S1–70.

34. Faintuch J, Matsuda M, Cruz ME, Silva MM, Teivelis MP, Garrido AB Jr, Gama-Rodrigues JJ. Severe protein calorie malnutrition after bariatric procedures. Obes Surg. 2004;14:175–81.

35. Parrott J, Frank L, Rabena R, Craggs-Dino L, Isom KA, Greiman L. American Society for Metabolic and Bariatric Surgery integrated health nutritional guidelines for the surgical weight loss patient 2016 update: micronutrients. Surg Obes Relat Dis. 2017;13:727–41.

36. Mechanick JI, Youdim A, Jones DB, Garvey WT, Hurley DL, McMahon MM, et al. Clinical practice guidelines for the perioperative nutritional, metabolic, and nonsurgical support of the bariatric surgery patient-2013 update: cosponsored by American Association of Clinical Endocrinologists, The Obesity Society, and American Society for Metabolic & Bariatric Surgery. Endocr Pract. 2013;19:337–72.

37. Heber D, Greenway FL, Kaplan LM, Livingston E, Salvador J, Still C. Endocrine and nutritional management of the post-bariatric surgery patient: an Endocrine Society clinical practice guideline. J Clin Endocrinol Metab. 2010;95:4823–43.

38. Papamargaritis D, Koukoulis G, Sioka E, Zachari E, Bargiota A, Zacharoulis D, Tzovaras G. Dumping symptoms and incidence of hypoglycaemia after provocation test at 6 and 12 months after laparoscopic sleeve gastrectomy. Obes Surg. 2012;22:1600–6.

39. Laurenius A, Olbers T, Naslund I, Karlsson J. Dumping syndrome following gastric bypass: validation of the dumping symptom rating scale. Obes Surg. 2013;23:740–55.

40. Banerjee A, Ding Y, Mikami DJ, Needleman BJ. The role of dumping syndrome in weight loss after gastric bypass surgery. Surg Endosc. 2013;27:1573–8.

41. Nielsen JB, Pedersen AM, Gribsholt SB, Svensson E, Richelsen B. Prevalence, severity, and predictors of symptoms of dumping and hypoglycemia after Roux-en-Y gastric bypass. Surg Obes Relat Dis. 2016;12:1562–8.

42. Papamargaritis D, Koukoulis G, Sioka E, et al. Dumping symptoms and incidence of hypoglycaemia after provocation test at 6 and 12 months after laparoscopic sleeve gastrectomy. Obes Surg. 2012;22:1600–6

43. Tzovaras G, Papamargaritis D, Sioka E, et al. Symptoms suggestive of dumping syndrome after provocation in patients after laparoscopic sleeve gastrectomy. Obes Surg. 2012;22:23–8.

Nutritional Support After Surgery of the Colon

10

Filippo Pucciani and Anna D'Eugenio

10.1 Introduction

The colon is the part of large bowel employed for storing, transporting, and expelling feces. This complex activity is due to the interaction between motor activity of the muscular wall, the absorption/secretion activity of the mucosa, and the microbiota activity which modifies the endoluminal content. The form and structure of the stool are the results of the influence of these different activities on the enteric material that arrives as liquid at the level of the ileocecal valve and that is discharged in solid state. Colonic surgery can obviously modify these activities resulting in altered motility, absorption/secretion, and microbiota composition, causing changes in the bowel stool, bowel transit, and defecation.

In order to be able to describe the influence of colonic surgery on bowel function, it is useful to first deal with topics such as the stool composition, colonic microbiome, colonic motility, and absorption/secretion of the colonic mucosa factors that are all influenced and/or modified by colonic resection. This preliminary description will be instrumental to understanding the nutritional needs of colon surgery patients.

F. Pucciani (✉)
Department of Experimental and Clinical Medicine, University of Florence, Florence, Italy
e-mail: filippo.pucciani@unifi.it

A. D'Eugenio
Consultant Nutritionist, Ambulatorio di Medicina Integrata, G. Bernabei Hospital, Ortona, Italy

© Springer Nature Switzerland AG 2019
D. F. Altomare, M. T. Rotelli (eds.), *Nutritional Support after Gastrointestinal Surgery*, https://doi.org/10.1007/978-3-030-16554-3_10

10.2   Colonic Physiology

10.2.1 Stool Composition

Stools are comprised of the solid or semisolid remains of ingested food that cannot be digested or absorbed in the small intestine but that have been decomposed and modified by colonic microbiota. On average, humans eliminate 128 g of fresh feces per person per day having a pH value of around 6.6 [1]. An average chemical analysis reveals that fresh human stool is approximately around 75% water and 25% solid: the solid fraction consists of 30–50% bacterial biomass, 10–25% protein or nitrogenous matter, 10–25% carbohydrate or undigested plant material, and 10–15% fat [1]. The remaining solids are composed of calcium and iron phosphates, intestinal secretions, small amounts of dried epithelial cells, and mucus [1]. The main factor affecting fecal mass is the host's fiber intake: the greater the presence of hydrated fibers, the more voluminous and soft are the feces. The form of the stool depends on the time it spends in the colon. The Bristol stool scale is designed to classify the form of human feces into seven categories and is also useful for understanding the relationship between colonic transit and stool form: slower transit results in more compact and hard feces while faster transit in soft, liquid composition [2].

The seven types of stool are:

1. Separate hard lumps, like nuts (hard to pass)
2. Sausage-shaped but lumpy
3. Like a sausage but with cracks on the surface
4. Like a sausage or snake, smooth, and soft
5. Soft blobs with clear-cut edges
6. Fluffy pieces with ragged edges, a mushy stool
7. Watery, no solid pieces. Entirely liquid

    Types 1 and 2 are frequently associated with chronic constipation. Types 3 and 4 are optimal, especially the latter, as these are the easiest to pass. Types 5–7 are associated with increasing tendency to diarrhea or urgency. Feces have physiological odor, which can vary according to the host's diet and health status. Volatile substances contain odorous compounds such as benzopyrrole, methyl sulfide, *hydrogen sulfide*, *methanethiol*, and *dimethyl sulfide* [3].

Stool volume, type, and content may be altered by colonic surgery.

10.2.2 Colonic Microbiome

The term *microbiome* describes the interaction among microbes, their genomes, and their human host environment. Important functions of the microbiome include the processing of a wide range of dietary plant polysaccharides, provision of colonization resistance, shaping of the immune system, and regulation of host signaling pathways [4].

The composition of the microbiome varies among individuals and within individuals at different times. The microbiome is alive, metabolically active, and highly dynamic in response to multiple environmental factors such as diet. The colonic microbiome is comprised of bacteria and other important components such as fungi and viruses: bacteria reach densities of 10^{11} cells per gram of luminal content. The fecal microbiome is a mix of luminal and mucosally adherent bacteria from the colon. Most studies on the gut microbiome focus on fecal microbiome due to the ease of sampling. Ninety percent of human stool is made up of *Bacteroides* and *Firmicutes*, whereas *Proteobacteria*, *Actinobacteria*, *Verrucomicrobia*, *Fusobacteria*, and *Cyanobacteria* account for lower percentages. Pregnancy, age, and antibiotics influence microbiome composition. Antibiotics have been found to alter the taxonomic, genomic, and functional capacity of the human gut microbiome: broad-spectrum antibiotics reduce bacterial diversity, select for resistant bacteria, and enable intrusion of pathogenic organisms [5]. Colonic resective surgery involves the perioperative administration of antibiotics, and it is easy to understand that there can be postoperative clinical repercussions such as infective colitis and diarrhea. In colonic surgery, the gut bacteria are still considered a putative pathogenic source that can lead to further infections. Thus, mechanical bowel preparation (MBP) is still used in order to decrease the bacterial load in addition to the traditional effect of the so-called bowel cleansing. This procedure has an obvious, immediate impact on the composition of the gut microbiome, but it also alters the quality and the production of the protective mucous layer, which can permit an eventual bacterial translocation. Few studies have investigated and described the impact that bowel preparation with or without oral antibiotics conducted prior to colorectal surgery may have on the mucosa-associated and luminal colonic microbiome. An increased level of *Proteobacteria* and *Bacteroides*, a significant reduction in *Lactobacillaceae*, an increase in *Enterobacteriaceae* abundance at the family level, and a drastic change in the ratio of Gram-positive to Gram-negative species are often found [6, 7].

Diet is the major influence on shaping the structure of the microbiome. It has been shown that some food components in the normal diet, such as resistant starch, oligosaccharides, and plant fibers, are incompletely absorbed in the small bowel and enter the colon where the microbiome is active. Bacteria, fungi, and viruses act on residues and modify them into products of different size and composition. The colon harbors large amount of gas and 400–1200 mL/day are released as flatulence. It has been proven that a diet rich in fermentable residues may induce instability in the microbial ecosystem, thereby increasing the hydrogen and methane produced by colonic bacteria with a resultant increase in flatulence [8].

10.2.3 Colonic Motility

The motility of the colon is designed to transport the endoluminal content by means of the interaction of different mechanisms: colonic wall contractions move the contents a few centimeters (*segmenting activity*), to facilitate the reabsorption of water and electrolytes, as well as activating the thrust and transport of the contents for several tens of centimeters (*mass movements*). The integration of these motor activities results in the complex performance of the storage, transport, and expelling of feces.

10.2.3.1 Segmenting Activity

Segmenting activity is expressed by rhythmic phasic contractions (RPCs) [9]. RPCs cause slow net distal propulsion with extensive mixing/turning over of luminal content; tonic contractions aid RPCs in their motor function.

Colonic segmenting activity is the effect of electromechanical coupling that results from the spontaneous periodic depolarizations of colonic circular smooth muscle cells, called *slow waves*; RPCs are mainly generated by circular smooth muscle cell contractions in response to slow-wave activity. The maximum frequencies of RPCs cannot exceed their respective slow-wave frequencies, and the slow waves alone regulate the timing and direction of propagation of short-duration RPCs. Smooth muscle cells and their slow-wave activity are influenced by enteric neurons. Two types of motor neurons project from the myenteric plexus to the circular muscle layer: the excitatory motor neurons and the inhibitory motor neurons. Some studies have proposed that intramuscular interstitial cells of Cajal (ICC-IMs) serve as obligatory intermediaries between the motor neurons and smooth muscle cells [10]. The concept is that the excitatory and inhibitory neurotransmitters of the motor neurons excite the ICC-IMs, which then project a composite signal to the smooth muscle cells to contract or not. Excitatory neurons ease muscular contraction, and the circumferential anatomic configuration of the excitatory axons explains the ring-like contraction that is the morphological expression of RPCs. The wall ring-like contractions trap fecal boluses in small chambers, in order to facilitate the contact of boluses with colonic mucosa for the reabsorption of water and electrolytes. RPCs match up with radiologic colonic haustra: they show a frequency gradient that is smaller in the right colon and becomes maximum in the sigma. The result is therefore the overall slowing of the ab-oral progression of fecal boluses, which slows total colonic transit time, whereby the feces are stopped in the sigma, ready to be transferred by a mass movement into the rectum for a bowel stool. The distal prominence and origin of these motor patterns raise the possibility of their serving as a braking mechanism, or a "rectosigmoid brake," to limit rectal filling [11].

10.2.3.2 Mass Movements

Mass movements are expressed by manometric giant migrating contractions (GMCs). The average frequency of GMCs in the human colon is about 6–10 per 24 h, mean amplitude about 115 mmHg, mean duration about 20 s, and they propagate distally for 20–45 cm at a speed of about 1 cm/s [12, 13]. GMCs support colonic distal propulsive activity, and most GMCs produce mass movements from the ascending colon to the transverse or descending colon. Only those GMCs that originate in the sigmoid colon or proximal to it and propagate up to the rectum produce the urge to defecate or the act of defecation. The frequency of GMCs exhibits diurnal variation: it is reduced by about 80% during sleep [14] and significantly increases upon awakening in the morning, sometimes inducing the urge to defecate which is then followed by defecation [15].

The ingestion of a meal ($\approx$1000 kcal) enhances colonic motor activity for about 2 h. The increase in colonic motor activity following a meal (*gastro-colonic reflex*)—especially in the morning when the sigmoid colon and rectum are likely full—is a primitive reflex to prod the colon to empty in preparation for the entry of new digesta: this reflex may include GMCs and corresponding mass movements, and if it involves the sigmoid district, it may propagate to the rectum and induce a bowel movement.

A number of observations and pharmacological experiments provide significant clues to the generation of GMCs. (1) Occurrence of GMCs increases after a meal or upon awakening in the morning, probably as an effect of hormonal and nervous stimulation. (2) S2 and S3 sacral nerve stimulation increases the frequency of GMCs twofold throughout the colon, and the central stimulation of GMCs probably works through parasympathetic outflow [16]. (3) The stimulation of mucosal nerve endings by bisacodyl, a mucosal irritant contact laxative, stimulates colonic GMCs via the enteric motor neurons that release ACh. (4) Last but not least, short-chain fatty acids, often resulting from the breakdown of undigested carbohydrates for fermentation by anaerobic bacteria in the colon, stimulate the release of 5-HT from enterochromaffin cells. 5-HT acts on 5-HT$_3$ receptors of vagal afferent neurons to stimulate the vago-vagal reflex. The efferent vagal nerves act through nicotinic receptors on enteric motor neurons to release ACh, which in turn stimulates GMCs [17].

10.2.4 Absorption/Secretion of the Colonic Mucosa

The colonic surface epithelium is a simple cuboidal epithelium that serves as a protective barrier for the endoluminal environment. It is composed of absorbent cells and goblet cells: the first manage colonic ion and water transport and the latter synthesize, store, and secrete mucous granules. Colonic mucosa is arranged in crypts, and its internal cellular composition is different from that of the surface epithelium. Colonic crypts contain immature and precursor cells as well as specialized endocrine cells and Paneth cells secreting lysozyme, epidermal growth factor, and glycoproteins. Colonic mucosa is involved in absorption processes with fluid and electrolyte exchanges. The colon is a major site for water absorption, and approximatively 90% of the water contained in ileal fluid arriving to the ileocecal valve is absorbed by the colon. Sodium is absorbed actively via sodium/potassium ATPase until 400 mEq/day and is accompanied passively by water. Finally chloride is absorbed actively via a chloride-bicarbonate exchange. Vice versa, potassium is actively secreted into the colonic lumen. Several forms of diarrheal disease involve the dysregulation of colonic water absorption and ion transporters, and an associated imbalance between secretory and absorptive functions of the colonic epithelium may take place.

10.3   Colonic Surgery

10.3.1 General Surgical Considerations

Colorectal resections are performed for a wide variety of conditions including tumors (benign and malignant), inflammatory bowel diseases, diverticular disease, and other benign diseases. Elective and emergency surgery involve the same technical principles: anastomoses and use of ostomies are well-defined, and their packaging is performed in the same manner regardless of timing of surgery. The extent of the colonic resection depends on the size of the mesentery to be removed, the nature of the disease (benign or malignant), and the location of the lesion. Laparoscopic surgery offers an effective surgical alternative to open surgery and is the most used technique in elective surgery, because of its minimal invasiveness. Types of colorectal resections are:

– Ileocecal resection
– Ascending colectomy
– Right colectomy
– Extended right colectomy
– Transverse colectomy
– Left colectomy
– Extended left colectomy
– Sigmoid colectomy
– Total colectomy
– Restorative proctocolectomy
– Anterior resection of the rectum
– Hartmann's procedure
– Abdominoperineal resection

In the setting of emergency surgery, the bowel is usually unprepared, and patient hemodynamic may be unstable. Colonic resection is often combined with a diverting ostomy (ileostomy or colostomy), and it is also appropriate if the colon appears to have vascular impairment or if the patient is hemodynamically unstable, underfed, or immunosuppressed. Laparoscopic and robotic surgeries are usually performed today with decreased postoperative pain, earlier return of bowel function, and shorter postoperative hospitalization instead of laparotomic surgery [18].

The protocol of enhanced recovery after surgery (ERAS) is often followed in the postoperative course of colorectal surgery [19]. Optimal pain control, prevention of fluid overload, and aggressive postoperative rehabilitation, including the early recovery of oral feeding and mobilization, are the main points of the protocol. The ERAS pathway reduces overall morbidity rates and shortens the length of hospital stay, without increasing readmission rates [20]. A significant reduction in nonsurgical complications is also evident, although no significant reduction in surgical complications has been demonstrated. Early postoperative oral intake is aimed at decreasing the patient's response to surgical stress and at preventing excessive catabolism: right colectomy and Hartmann reversal constitute risk factors for having delayed tolerance to normal diet [21]. Tolerance to early oral feeding is not influenced by postoperative ileus in a randomized controlled trial [22]: another study also shows accelerated gastrointestinal recovery by early resumption of oral diet, resulting in promotion of gas passage and defecation [23].

10.3.2 Introduction to Colonic Surgery for Functional Diseases

Colonic surgery is used in colonic functional diseases only when conservative management, including rehabilitative treatment, has failed.

The same premises regarding colonic physiology and the influence of surgery must be known if a colonic resection is planned for a colonic functional disease. The only difference is the absence of lymphadenectomy, a technique that is used in colonic malignant neoplasms surgery.

Colonic surgery for functional disease is aimed at treating slow-transit constipation caused by colonic inertia. The surgical procedures include the following:

1. Total or subtotal colectomy with ileorectal anastomosis.
2. Subtotal colectomy with antiperistaltic cecorectal anastomosis.
3. Malone antegrade continence enema (MACE).

There are no randomized trials involving the above three procedures.

10.3.3 Colectomy

Total or subtotal colectomy with ileorectal anastomosis is performed after failure of behavioral, dietetic, pharmacological, and rehabilitative treatment to improve the symptoms [24–26]. Selection criteria of patients include the following symptoms [27]:

(a) ≤2 weekly defecations
(b) Duration of symptoms (>5 years)
(c) The presence of symptoms such as abdominal bloating or pain, nausea, and vomiting that have a significant impact on the patient's quality of life
(d) Failure of conservative therapy
(e) Radiological evidence of slow-transit constipation
(f) Exclusion of organic or functional pelvic floor disorders
(g) Exclusion of upper gastrointestinal tract dysmotility
(h) Normal psychological evaluation

The overall rate of success or satisfaction is high: in 39 studies involving 1423 patients, the success rate was 86%; the rate of bowel movements reported by the patients was 2.8 times per day, whereas recurrent constipation was reported by 9% [27]. Urgency and urge incontinence may appear in about 15% of patients, whereas small intestinal obstruction was sometimes reported. There are not randomized trials to test whether total colectomy (with ileorectal anastomosis) is better than subtotal colectomy (with ileosigmoid anastomosis), but one study indicated that there is no difference in the patient's postoperative general condition between these surgical techniques [28]. Feng et al. reported that subtotal colectomy can significantly normalize the number of intestinal flora by changing preoperative fecal and colonic mucosal microbiota [29].

Laparoscopic total colectomy has been proven to be safe and increases the number of evacuations per week, thus improving the patient's quality of life [26]. One comparison between total colectomy for colonic inertia (TCCI) and for noninflammatory indications (TCNI) shows that the TCCI group had a significantly higher incidence of postoperative ileus, higher overall morbidity, and significantly increased length of hospital stay [30]. These differences persisted when subgroups of patients who underwent laparoscopic or open surgery were compared. There are no explanations for these findings.

10.3.4 Subtotal Colectomy with Antiperistaltic Cecorectal Anastomosis (Figs. 10.1 and 10.2)

The subtotal colectomy with isoperistaltic cecorectal anastomosis was first proposed by Lillehei [31]. His technique was modified and today antiperistaltic anastomosis is performed [32]. Cecal stump and antiperistaltic anastomosis both contribute to

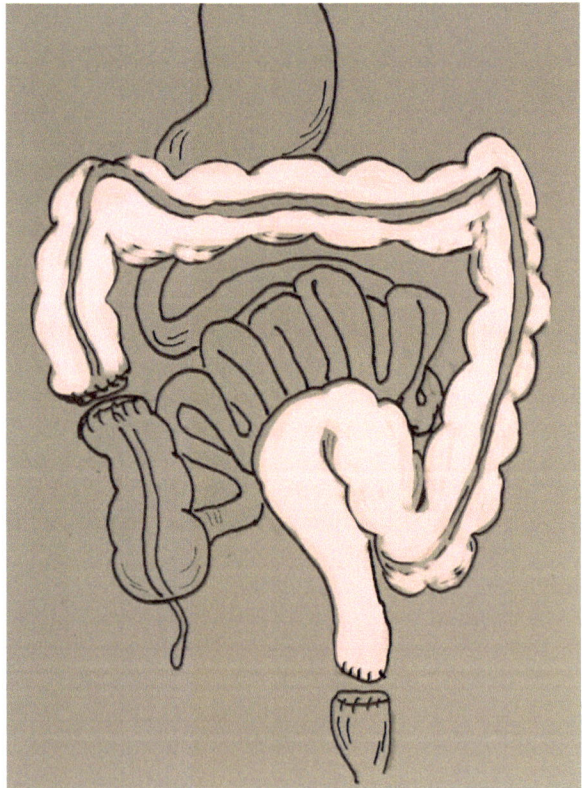

Fig. 10.1 Subtotal colectomy (in pink the colonic segments to be removed)

Fig. 10.2 Antiperistaltic cecorectostomy

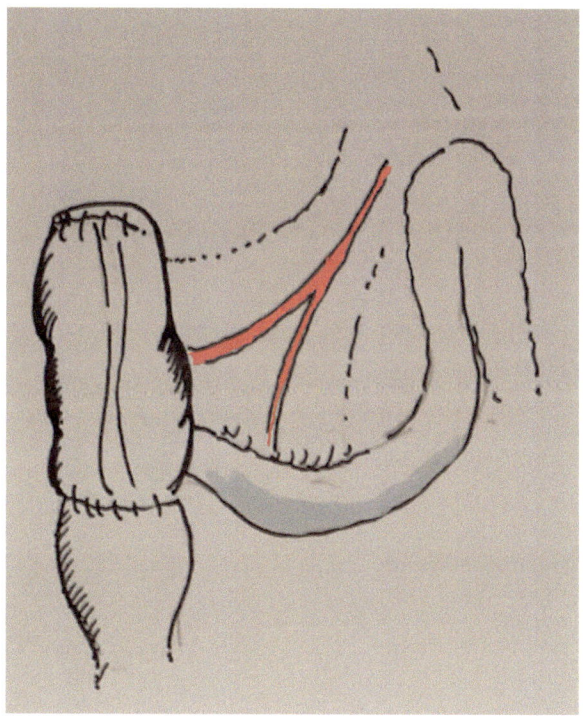

slowing the influx of enteric fluid into the residual rectum, thus slowing ileoanal transit. The cecal stump, on the one hand, acts as a "reservoir" by increasing rectal capacity; on the other, it allows reabsorption of water and electrolytes decreasing the amount of liquid flowing into the rectum. The ileocecal valve normally acts as part of the "ileal brake" and allows an orderly progression of contents into the cecum, also preventing the backflow of the cecal content and thus avoiding potential colonic bacterial contamination of the small bowel. All these mechanisms contribute to better continence after total colectomy. The cecal stump has a maximum length of 8 cm, and the cecal "reservoir" should not give rise to any long-lasting discomfort such as diarrhea, incontinence, or soiling [33]. Long-term follow-up of at least 60 months in patients shows a mean frequency of defecation of 0.9 ± 0.5 times per day and patient satisfaction about 82.1% [34]. However, there are no randomized trials comparing total colectomy with ileorectal anastomosis and subtotal colectomy with antiperistaltic cecorectal anastomosis.

10.3.5 Malone Antegrade Continence Enema (MACE)
   (Figs. 10.3 and 10.4)

Malone antegrade continence enema was firstly used for fecal incontinence [35]. The original procedure combined the principle of an antegrade colonic enema for

Fig. 10.3 Schematic representation of an antegrade colonic enema procedure (Malone)

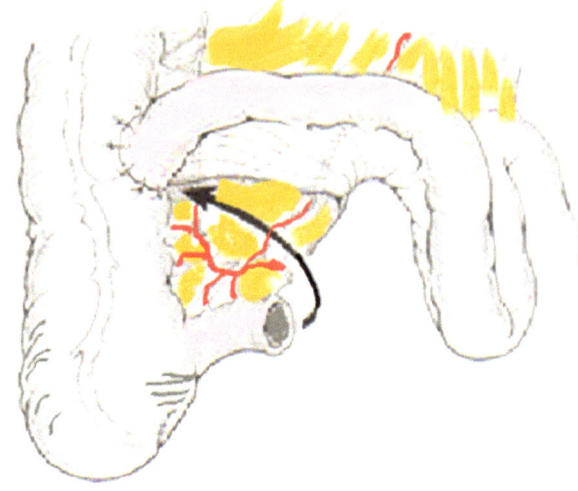

Fig. 10.4 Schematic representation of an antegrade colonic enema procedure (Malone)

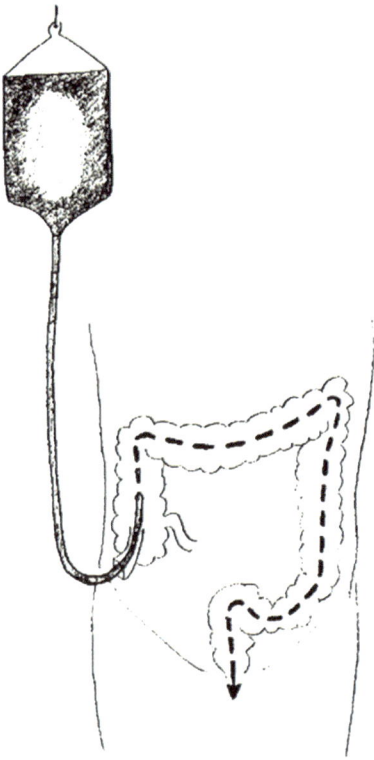

colonic washout with the Mitrofanoff technique of forming a continent appendicostomy conduit that could easily be catheterized via the abdominal wall. The Malone procedure was subsequently introduced for treatment of severe idiopathic constipation, even in pediatric age [36, 37]. There have been numerous technical variations with the latest being a small, introflexed ileostomy modified by means of the Marsh and Kiff technique [38]. To date, postoperative irrigation regimens (enema type, volume, and frequency) must be individualized to each patient. The MACE effectively treats intractable constipation that has been unresponsive to medical treatment. The results of this procedure have been very satisfactory, although study follow-ups are all short term. The overall success or satisfaction rate, documented in 7 studies on 67 patients, is 74%, but, within 3 years, the MACE procedure has been replaced by other therapies in 50–75% of cases [27]. There are no randomized or controlled studies on MACE in the literature, but this technique could be used before embarking on the path of total colectomy.

10.4   Effects of Surgery on Colonic Physiology

Colorectal resections may influence colonic physiology and function depending on type of surgical operation, site, and relative length of resection. In any case, colonic resection causes a shortening of the colon: in this way the speed of the total colonic transit increases, and therefore total transit times are reduced.

Several colorectal operations will be evaluated in detail below, to improve understanding of this topic.

10.4.1 Ileocecal Resection: Right Hemicolectomy
(Figs. 10.5 and 10.6)

Both types of operations remove the ileocecal valve, changing the enteric liquid inflow into the colon. The valve presumably contributes to gut homeostasis by optimizing retention of chyme in the small intestine until digestion is largely complete; the ileal contents should then be programmed to empty into the large bowel in a manner which does not compromise the colon's capacity to absorb [39]. In healthy humans, the motor responses of the ileocecal valve are consistent with a sphincteric function: basal tone is augmented by cecal distension or a meal and is inhibited by ileal propagating pressure waves [40]. This combined motor activity prevents reflux from the cecum into the ileum and acts as an intake valve of enteric material into the cecum. Removal of the valve causes uncontrolled and continuous flow of the enteric fluid into the colon: more than 1500 mL/day can enter the bowel downstream by overloading the residual colon with water and electrolytes. The overload will be absolute after right colectomy, especially if extended, or relative in cases with ileocecal resections: the cecum, right colon, and transverse colon are in fact the colonic sections where water reabsorption is highest when compared to the other colonic sections. Ileocecal valve removal also allows easy passage of colonic bacteria into the ileum: this phenomenon may induce small intestinal bacterial overgrowth (SIBO) syndrome that is characterized by diarrhea and weight loss [41]. Fast transit time, liquid overload, and bacterial overgrowth, all together, explain the possible high frequency of bowel movements that may appear after ileocecal resections or right colectomies.

Fig. 10.5 Schematic
representation of a right
colectomy and ileocolon
anastomosis

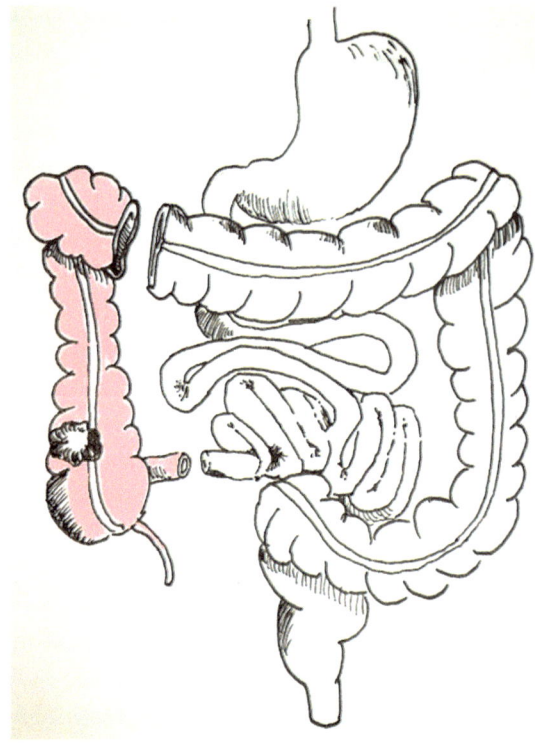

Fig. 10.6 Schematic
representation of a right
colectomy and ileocolon
anastomosis

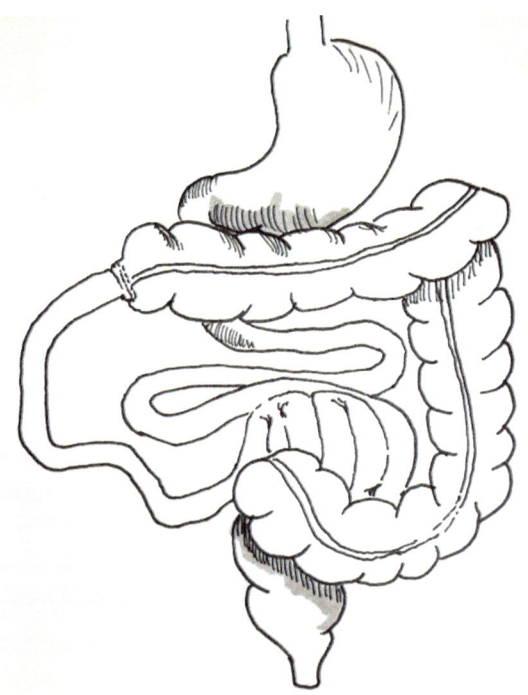

As previously mentioned, the ileocecal valve affects small bowel motility: by regulating the passage of chyme into the colon, it is part of the "*ileal brake*" that is the primary inhibitory feedback mechanism which controls transit of a meal through the gastrointestinal tract, thus optimizing nutrient digestion and absorption [42]. The absence of the ileocecal valve may impair the "ileal brake" disarticulating the motor coordination of intestinal loops. Postoperative ileus may be prolonged due to small bowel dysmotility: gastric stagnation could last for a few days until coordinated gastric and small bowel motility restarts.

10.4.2 Left Colectomy (Figs. 10.7 and 10.8): Sigmoid Colectomy

Regardless of the type of surgery involving a sigmoid resection (left colectomy, extended left colectomy, sigmoid colectomy, total colectomy, anterior resection of the rectum, and recanalization after Hartmann's procedure), the functional effects on colonic physiology will always be the same. The removal of the "rectosigmoid brake" involves, in addition to the variable shortening of the bowel—depending on the type of surgical intervention—a fast colonic transit with decreased colonic transit time [43]. The result is increased defecatory frequency compared to the

Fig. 10.7 Schematic representation of a left colectomy

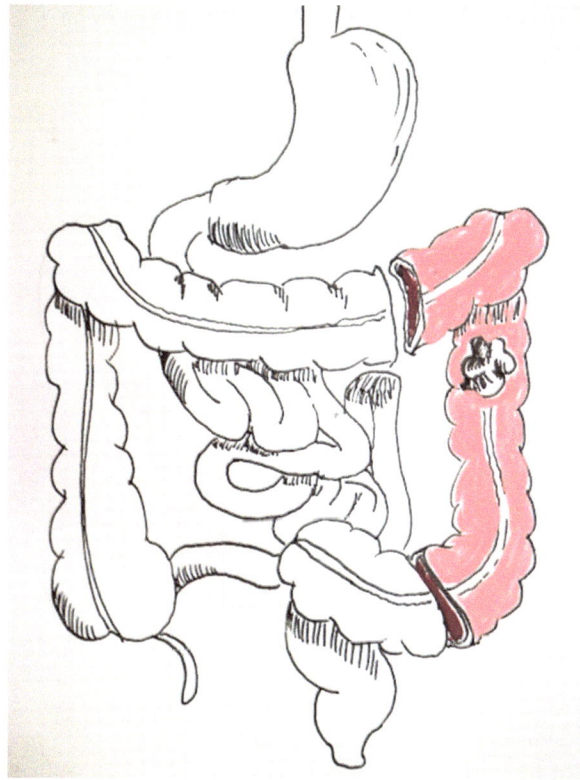

Fig. 10.8 Schematic representation of a left colectomy and colorectal anastomosis

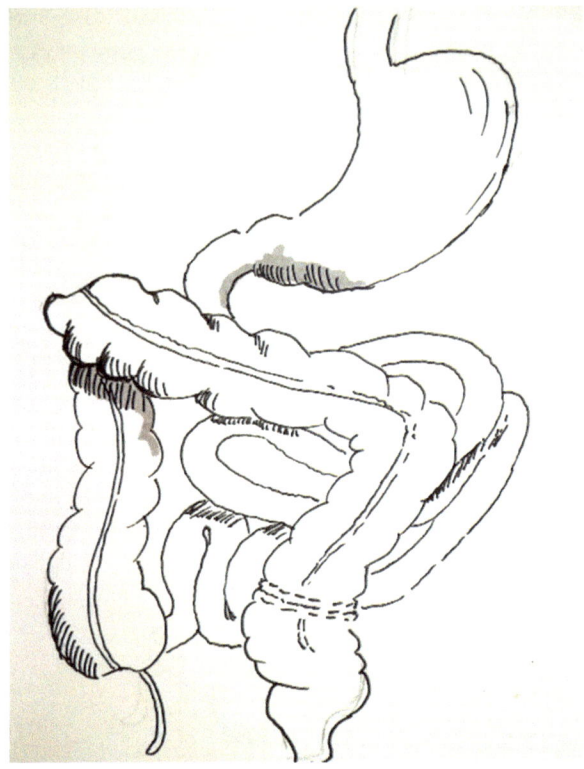

preoperative situation with many bowel movements per day of loose stools. Colonic manometry may reveal that the background contractile segmental activity is heavily reduced, sometimes combined with increased frequency of giant migrating waves [44]. Manometry suggests that the full colonic propulsive activity acts against a resistance decreased by the impaired segmental activity.

10.4.3 Anterior Resection of the Rectum (Figs. 10.9 and 10.10)

Anterior resection of the rectum, especially if performed with ultralow colorectal anastomosis, coloanal anastomosis, or intersphincteric resection, besides showing the effects of sigmoid resection on colonic transit, may have negative influences on bowel movements, which are often worsened after adjuvant or neoadjuvant radiotherapy, and sometimes result in low anterior resection syndrome (LARS) [45]. The backdrop of functional asset for these surgical operations is a combination of a small neorectal capacity and related high endo-neorectal pressures which together act against a weakened sphincteric mechanism [46]. After rectal resection a colonic segment is mobilized and anastomosed to the rectal stump or anal canal to restore intestinal continuity. Thus, a colonic loop, used to substitute the resected rectum, is placed to perform inappropriate functions with the result that colonic motility, sensitive

Fig. 10.9 Schematic representation of an anterior resection of the rectum

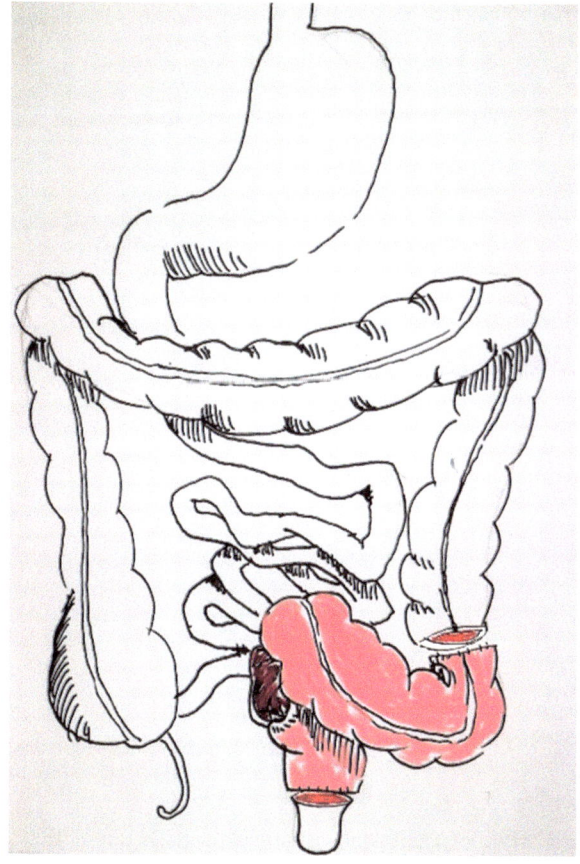

perception, and tonic accommodation to contents are all quite different from those of the rectum. The neorectum is also smaller than the rectal ampulla implying that the volumetric capacity is reduced and that only small fecal amounts may be stored which are not immediately evacuated. This functional aspect has led to the colonic J-pouch as reconstructive method in order to provide a better volumetric capacity. Effectively the colonic J-pouch reduces daily stool frequency and urgency in comparison with a straight coloanal anastomosis, but, after 2 years, this difference tends to disappear [47]. Beyond the impairments of the neorectal stump, there are also several negative influences on anal sphincter function. Anal sphincter damage may occur, and it is mainly due to injuries of the internal anal sphincter: up to 18% of patients who undergo stapled low anterior resection have had long-term evidence of internal anal sphincter injury [48]. Internal anal sphincter function may also be compromised if the internal anal sphincter nerves are damaged [49, 50]. Finally, ultralow anterior resections may sometimes be associated with anal sensation defects, which sometimes occur after intersphincteric resection. Excisions of anal mucosa, particularly in the anal transitional zone, might destroy the typifying receptors that allow the patient to discriminate between flatus and liquid or solid stool.

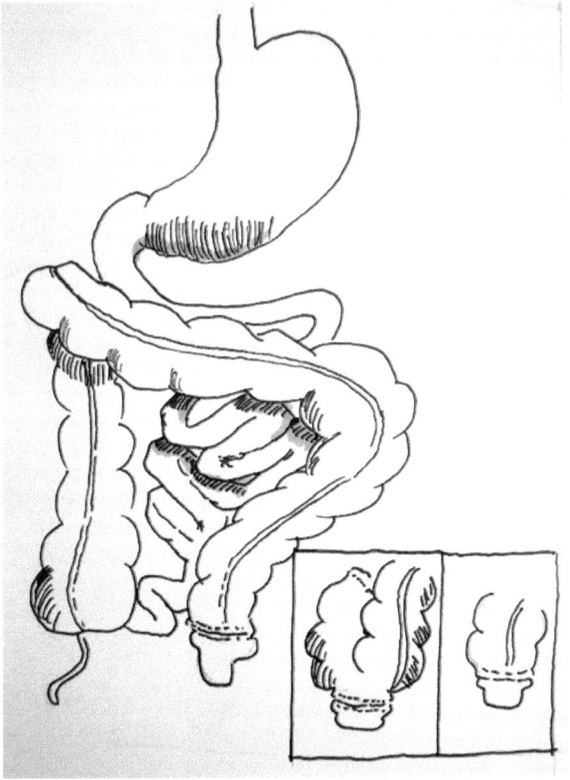

Fig. 10.10 Schematic representation of an anterior resection of the rectum and coloanal anastomosis

All these effects on colonic capacity and anal function, variously combined, are the main reasons that lead to the great variety of signs and symptoms in ARS: they include a mix of high bowel frequency/day with liquid stools, sometimes multiple evacuations with multiple movements within a limited time period, urgency, and fecal incontinence.

10.4.4 Total Colectomy: Restorative Proctocolectomy
(Figs. 10.11 and 10.12)

In total colectomy and ileorectal anastomosis or in restorative proctocolectomy with ileal pouch-anal anastomosis, after colectomy with or without rectum removal, intestinal continuity is restored by using the ileum. The ileum is a bowel that has no volumetric containment capacity, nor even content storage. In addition, there are two other important factors: first, ileal content is liquid, and, second, the transport of intraluminal fluid is constant until it reaches the anastomotic site. Even if the rectum has been saved or a pouch is created to implement storage, ileal motility and content fluid type are not much influenced, thus establishing the conditions for high defecatory frequency. Manometric studies in restorative proctocolectomy show that

Fig. 10.11 Total proctocolectomy (in pink the organ removed)

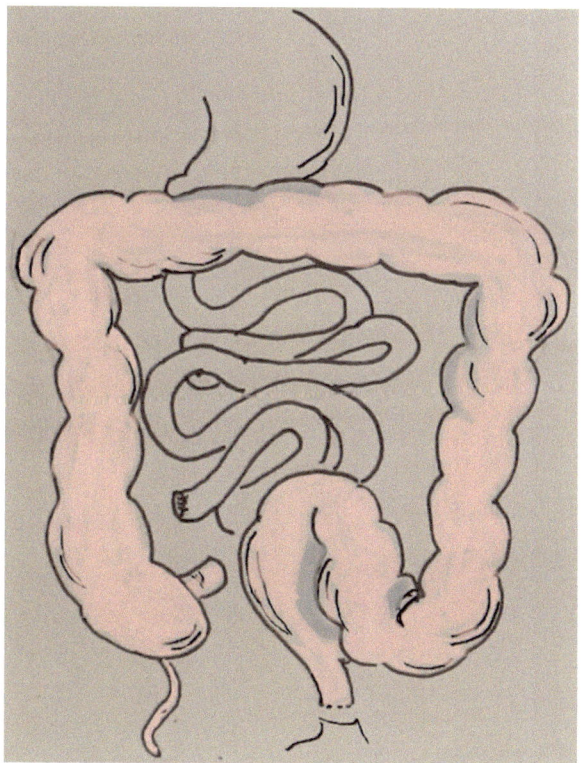

overall motility is reduced in the pouch when compared with the distal ileum and that propulsion, identified by large propulsive waves, in the pouch loop system is opposed by the activity of the anal sphincters [51]. This report supports the motor model of pouch storage activity but explains the continuous ileal incoming fluid.

Inflammation of the ileal pouch (pouchitis) is a common complication. Although definitive proof is still lacking [52], clinical practice indicates that mucosal inflammation is localized in the area of the gut characterized by the highest bacterial concentrations and therefore that short-term oral antibiotic therapy (metronidazole or ciprofloxacin) is an effective treatment [53]. This supports the hypothesis that a microbial imbalance of the pouch microbiota plays an important role in the pathogenesis of pouchitis. Although a particular bacterial species associated with pouchitis has not been identified, *Enterococcaceae* spp., *Enterobacteriaceae* spp., and *Streptococcaceae* spp. may play active roles in maintaining immunologic homeostasis within the pouch mucosa, as low levels seem to favor the development of acute pouchitis [52].

Urgency and fecal incontinence may appear, with malfunctioning of the neorectum and/or anus. Preservation of the rectum positively influences intestinal function. In patients affected by familiar adenomatous polyposis, total colectomy with ileorectal anastomosis has been compared with restorative proctocolectomy with ileal pouch-anal anastomosis [54]. Bowel frequency, night defecation, and the use

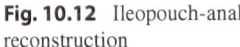

Fig. 10.12 Ileopouch-anal reconstruction

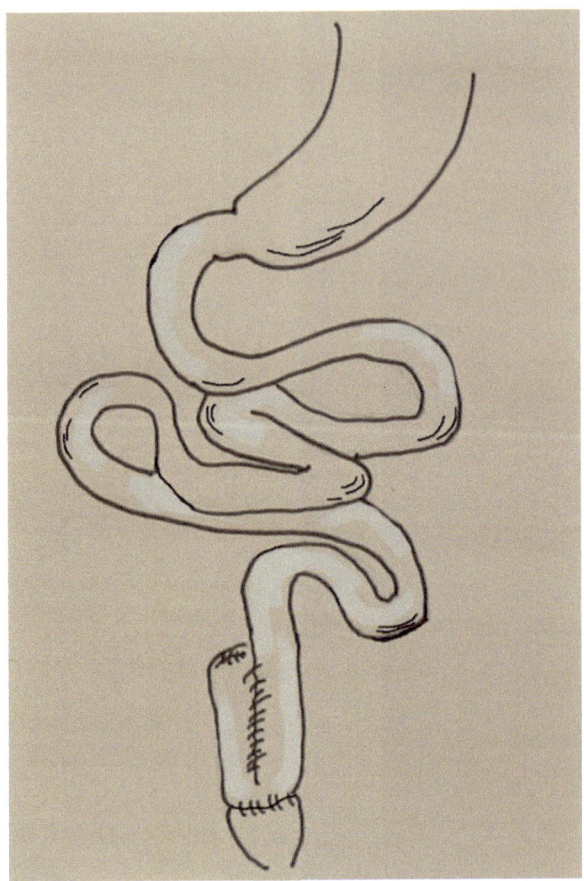

of incontinence pads were significantly less in the ileorectal group, although fecal urgency was reduced with the ileal pouch. This implies that rectal volumetric capacity and rectoanal coordination are determinant for an optimal defecation.

Laparoscopic total colectomy is often performed, since the procedure is as safe as the open technique. Long-term defecatory function is better following laparoscopic total colectomy than following open total colectomy: the mechanism for this improvement is unclear, but it might relate to the underlying reason for surgery or possibly to reduced small bowel handling leading to fewer adhesions after laparoscopic surgery [55].

10.5 Conclusions

Colorectal resections influence the storage, transport, and expulsion of feces. Surgery can modify bowel movements, sometimes resulting in impaired defecation. In the most serious cases, a disabling physiological condition can arise which has a negative psychological impact on the patient and results in a lower quality of life.

10.6   Nutritional Support After Surgery on the Colon and Rectum

10.6.1 Nutritional Support After Segmental Colectomy

With the surgery and the consequential removal of the neoplastic mass, the body has a better chance of recovery [56], but there is often a weight loss which adds to what has already been lost because of the illness, and so the immune system is weakened [57, 58].

Getting back to eating properly is the best way to recover. Many hospitals adopt the ERAS program (enhanced recovery after surgery), a program born with the aim of minimizing the metabolic, neuroendocrine, and whole organism response to the surgical stress, obtaining an earlier functional recovery and a reduction of postoperative complications. As a result, there is an optimal recovery and an early and safe return to daily activities.

The nutrition of the patient discharged after resection of the colon must bear in mind the need to reintegrate nutrients in order to have a good recovery of strength but also the limits imposed by the new anatomy, which will influence in various ways the intestinal movement. Energy needs to return to being that of a healthy person, but digestive possibilities are partially diminished [59], mainly those linked to the intestinal transit but also those linked to the capacity to produce certain vitamins [60].

Surgical resection of the intestine brings about a change in the intestinal microbiota [61]; it causes in turn alterations in transit, with a greater tendency to diarrhea; however, constipation is the complication to be feared most.

Acting on the microbiota is the main purpose of the diet, because this is able both to ensure a good transit and to interact with the immune system [62], acting positively on the evolution of the disease. Scientific literature has many works that confirm the positive role of a diet prevalently rich in fiber (vegetables, wholemeal cereals, and pulse), with a low protein content, able to favor the growth of different kinds of bacteria with an anti-inflammatory action [63, 64], but the limits imposed by the recent surgery point in the initial phase toward a low fiber diet, to allow the anastomosis to heal completely and the intestine to go back to tolerating it better. The aim however is to enable the organism to tolerate a diet with a good amount of vegetables again. Intestinal bacteria in fact during the digestion of fruit and vegetables produce short-chain fatty acids (SCFA), with a well-recognized antitumoral action [65].

In the first phase, the diet will be based on vegetable broth, first without creamed vegetables, with the addition of rice or small pasta. Two weeks after the surgery, creamed vegetables can begin to be reintroduced progressively. The daily amount of protein will be satisfied by eggs, fish, and lastly meat. Freshly juiced fruit and vegetables will give plenty of vitamins and mineral salts without a fiber load. It is better not to extract juice only from fruit to avoid glycemic and insulinemic peaks, particularly harmful in the case of neoplastic pathology [66].

The introduction of foodstuffs must be progressive; bread and pasta should be introduced with caution because they are a possible cause of abdominal swelling. Intestinal dysbiosis, following surgery, is always accompanied by damage to the intestinal barrier with a resulting increased permeability of the intestine. It is important to try to eat foods capable of repairing this barrier and avoiding or reducing those with an inflammatory action [67]. Foods like yoghurt are useful for their content of live cultures, possibly divided up into small amounts during the day, and ginger and curcuma [68] which have a proven anti-inflammatory action.

One month after surgery, it is possible to eat all kinds of food remembering, however, that:

- While useful, fiber can create problems of swelling and abnormal transit, so pay attention to pulse (chickpeas, beans, lentils, peas) which must always be creamed.
- Prefer semi-refined cereals and not whole grain.
- Green leaves (chicory, beets, artichokes, edible thistles, spinach, celery) should be consumed moderately, not more than once a day.
- Vegetables which produce a lot of gas such as cruciferous plants (broccoli, cabbage, spring greens, black cabbage) or the garlic family (garlic, onions, leeks) should be avoided at first and then used with moderation.
- Refined white flour (0 and 00) should be excluded.
- White sugar should not be used, brown sugar greatly reduced, and sweeteners avoided.
- Great attention should be paid to yeast, likely to cause notable abdominal swelling; bread should always be leavened with sourdough. In any case it is a good rule to toast bread lightly before eating to make it more digestible.

Intestinal dysbiosis is very frequent in people who have undergone surgery for colon cancer. Some authors have found an increase of Firmicutes bacteria; others have identified some microbial stocks of the *Bacteroidetes* and *Prevotella* families. The diet must not be too high in fats and animal proteins and too poor in vegetable fibers because this would create an enteropathogen microbiota, with a reduction in the concentration of short-chain fatty acids in the feces, in particular of Butyrate acid. The action of this fatty acid as an inhibitor of cellular proliferation and of angiogenesis, and as inductor of apoptosis, makes it particularly useful to reduce the risk of relapses [69].

The synthesis of bacteriocins by an efficient microbiota, besides, limits and prevents a proliferation above all of the *Clostridium* and *E. coli*, fearful causes of diarrhea.

Many foods (chicory, artichokes) contain prebiotic substances, useful for the growth of intestinal bacteria, but their consumption must be limited because of the large amount of fiber residue which can, in case of postoperative adherences, increase sub-occlusive risk.

To improve the quality of the intestinal environment, it is necessary to improve digestion, because this determines which and how many nutrients more or less badly digested will reach the colon [70]. A diet with a small number of foods and of poor variety according to the cooking method does not carry out this task. Digestive processes are activated by smells as well as flavors, and apart from barbecue grilling, dangerous for its production of heterocyclic amines [71], all others can be used. Also frying, as long as it is done with extra-virgin olive oil and as long as the oil does not reach high temperatures, can be used occasionally (better rarely) [72], always associated with raw vegetables (rich in antioxidant substances).

10.6.1.1 Harmful Foods

- All preserved foods because of the large amounts of additives, in particular cured meats, full of nitrites and nitrates proven to be carcinogenic for the digestive system.
- Sweets, both for their stimulating effect on the production of insulin and the consequent proliferation and for their inflammatory effect.
- Sweet and fizzy drinks.
- Alcohol, because of its inflammatory action which reduces the capacity to repair damaged cells and for acetaldehyde, the substance which alcohol is changed into and which is recognized as carcinogenic [73]. Alcohol reduces the capacity of folate absorption which are protective compounds for colon cancer.

10.6.1.2 Useful Foods

- Freshly extracted fruit and vegetable juices, remembering to use a good amount of vegetables to avoid glycemic spikes. These drinks have the advantage of bringing an important number of nutrients to the organism, linked above all to the vegetables, which a person who has undergone an operation for colon carcinoma cannot eat freely. Using a fresh juice of cabbage, carrot, apple, and ginger, for example, it is possible to introduce sulforaphane as well as indole-3-carbinol, very important substances with the antitumoral action of cabbage [74]; beta-carotene and vitamin C of the carrot [75]; and pectin of the apple. The gingerols and shogaols of ginger help digestion, reduce nausea, and contribute to improving the intestine.
- Green tea, rich in polyphenols, in particular epigallocatechin gallate with its powerful antioxidant action is able to block angiogenesis [76].

Food intake changes in case of feces tending to diarrhea or constipation.

In the first case, fiber should be reduced, using creamed cereals (e.g., semi-refined barley cooked in vegetable broth and then sieved to obtain a cream), steamed meat, or fish. Freshly extracted juices can be drunk during the meal instead of water. It is useful to eat parmesan cheese, rich in protein and mineral salts, to reintegrate the loss of electrolytes and the resulting elimination of bile in the intestine, with improvement of peristalsis.

In case of greater intestinal irregularity, without daily evacuation, a day of liquid diet can be alternated regularly: the lack of residue will facilitate a return to normal bowel function.

10.6.2 Nutritional Support After Anterior Resection of the Rectum

Irregularity in the pattern of bowel movements and the different way of evacuating, together with poor continence, are the symptoms which the patient with anterior resection of the rectum has to face [77].

The main causes are the reduced capacity of the rectal ampulla together with neurological alterations which modify both the tone of the sphincter and the inhibitor reflex of the anus.

Diet becomes greatly influenced by the new anatomy, and great attention must be made to limiting foods which create flatulence, increase of fecal loss, and unpleasant smells [69] (particularly garlic and onions, cabbage, artichokes, Jerusalem artichokes, pulse, and kiwis).

It is very important to control body weight. A rapid weight loss can lead to a loss of lean body mass and with this a further reduction in the containing capacity of the sphincter. A weight increase should also be avoided because it can change the functioning of the stoma [78], but it can also create inflammation with a worsening of the intestinal environment [79].

Changes in digestive capacity, as well as of the bowel microbiota, strongly influence the quality and quantity of the fecal mass and its transition [80]. It is important to concentrate on all the phases of digestion, trying to improve them so as to reduce the negative repercussions on evacuation to a minimum.

Thorough chewing of the food, aided by the production of saliva, starts digestion. Foods able to create a good saliva output are those with a bitter flavor (endives, wild chicory, dandelion, trevisano radicchio) and lemon juice. At a gastric level, it is important to avoid medicines that contrast acidity (anti-acids, inhibitors of the protonic pump) because without the correct acid pH, the stomach is not able to digest, and so poorly digested foods reach the bowel.

In the small intestine, where these collect once they have left the stomach, the digestive processes are completed, and absorption begins. The intestinal bacteria contribute to this mechanism, above all the Bifidus bacteria which maintain the efficacy of the narrow junctions of the cell membranes [81], eliminate the toxic substances produced or swallowed, and decide the use of food residue preparing it for saprophyte bacteria, produce lactic, butyric, and propionic acids able to maintain an environment inhospitable for the reproduction of germs [82].

It is only in this way, supporting the digestive processes and making available to the intestinal flora the energetic sublayers useful to maintain them, that risk of infections can be reduced, which can happen easily when the anatomy of a body region changes.

Vegetables can be used alone, always chopped very finely or creamed. Zucchini and carrot, pumpkin, zucchini, and potato soups are useful. Creaming vegetables allow also cabbage to be eaten, usually problematic, both because it increases intestinal peristalsis and because it produces gas. Creamed soups of cauliflower or cabbage and potatoes, of black cabbage, and of pumpkin can be tried, starting from small amounts preferably mixed with boiled rice; basmati is excellent.

Meat and fish do not create particular problems; red meat should be consumed moderately, once or at most twice a week in case of anemia.

Eggs are well tolerated if cooked for a few minutes; hard-boiled eggs are to be avoided, much more indigestible and the cause of smelly gas.

Foods difficult to deal with are milk, cheese, and pasta.

The alteration of the intestinal microbiota following the pathology and the various types of surgical operation necessary to remove the tumor is accompanied by damage to the intestinal mucosa which becomes much more permeable [83]. The result is a different capacity to digest lactose and gluten. When these are eaten, symptoms such as slow digestion and gastric and abdominal swelling, meteorism, diarrhea, or constipation occur. Seasoned cheeses containing much less lactose are better tolerated but tend to produce more solid feces. In my personal experience, it is always better to substitute wheat pasta with pasta made of spelt or barley alternated with pasta made of gluten-free cereals (rice, millet, buckwheat, quinoa).

Alternating food stuffs, their associations and their cooking methods bring about improved digestion and nutritional quality, but it is fundamental to concentrate on the quality of the raw material. Vegetables must be cultivated without pesticides, animal proteins must come from correctly bred cattle, when the fewest possible drugs are used.

Use preserved foods as little as possible. The strength of food intake, its nutritive, and repairing capacity are linked to the possibility of receiving energy from the food, an energy directly in proportion to its freshness.

10.6.3 Nutritional Support After Proctocolectomy

The new ileus-anal reservoir, able to create a containing cavity for the feces in substitution of the rectum removed, allows the patient to have a good quality of life but does not prevent a series of functional changes. The lack of the colon causes an alteration of the feces because the organ which reabsorbs 90% of the liquid daily present in the feces is lacking. There will therefore be frequent and loose evacuations [84].

The most frequent clinical situations to be corrected are dehydration, the loss of electrolytes, vitamin deficit, and hypoalbuminemia.

To avoid hydro-saline imbalance, it is important to drink, but it may be useful to make up a solution of water, the juice of one or two lemons, a little cane sugar, and a pinch of salt. In this way a poly-saline drink, able to reintegrate the electrolyte losses, is created, well tolerated at a gastric level [85].

When daily evacuation is frequent, meals should be made up mainly of dry food-stuffs, pasta or rice, toast, and animal protein. Rice can be cooked in vegetable broth so as to make a risotto, therefore a dry dish with a high mineral salt content.

The use of eggs is important, above all egg whites, to avoid protein impoverishment. Parmesan cheese is also useful or in general matured cheese rather than soft cheeses.

A deficit of vitamin B12 and iron will be compensated by the daily use of meat juices, particularly useful in the form of juice of beef diaphragm. This part of the animal contains 2.8 mg/100 g of iron and 3.7 μg/100 g of vitamin B12, more than any other part. It is sufficient to cut 100 g into pieces, put them in a glass container, and cook it in a double boiler until the juice comes out. This juice can be added to any food or drunk with a little lemon juice to facilitate absorption. In this way normal levels can be raised without creating digestive problems that meat may give, above all in the cancer patient [86].

When the number of evacuations is reduced, more foodstuffs can be introduced, including vegetables. Carrots, courgettes, pumpkin, radicchio leaves, curly endive, and Belgian endive can be safely used; in the case of vegetables like spinach and Swiss chard, it is better to chop them into small pieces to make them more easily digestible. Artichokes, Jerusalem artichokes, aubergines, mushrooms, cabbage, garlic, and onion should be eaten very cautiously above all for their meteoric effect. Milk and dairy products can cause diarrhea, along with sweet pastries.

Foods which can cause inflammation (above all sweets, cured meats, and all foods full of additives) should be avoided as far as possible, not only for fear of increasing the number of evacuations but above all to protect from the risk of pouchitis, the worst complication [87]. Clinically this presents itself with an urgent need to defecate, an increase in the number of evacuations, and diarrhea, usually typically with bloody mucus. The most likely hypothesis about etiopathogenesis is of the inflammatory type, caused by excessive growth of pathogen bacteria and of the ischemic microcirculatory type. In both cases sugar acts negatively as it encourages micro growth, and through the insulin stimulus, it facilitates oxidative stress and the production of thrombotic factors.

References

1. Rose C, Parker A, Jefferson B, Cartmell E. The characterization of feces and urine: a review of the literature to inform advanced treatment technology. Crit Rev Environ Sci Technol. 2015;45:1827–79.
2. Lewis SJ, Heaton KW. Stool form scale as a useful guide to intestinal transit time. Scand J Gastroenterol. 1997;32:920–4.
3. Tangerman A. Measurement and biological significance of the volatile sulfur compounds hydrogen sulfide, methanethiol and dimethyl sulfide in various biological matrices. J Chromatogr B Analyt Technol Biomed Life Sci. 2009;877:3366–77.
4. Hooper LV, Gordon JI. Glycans as legislators of host-microbial interactions: spanning the spectrum from symbiosis to pathogenicity. Glycobiology. 2001;11:1R–10R.
5. Modi SR, Collins JJ, Relman DA. Antibiotics and the gut microbiota. J Clin Invest. 2014;124:4212–8.
6. Bucher P, Gervaz P, Egger JF, et al. Morphologic alterations associated with mechanical bowel preparation before elective colorectal surgery: a randomized trial. Dis Colon Rectum. 2006;49:109–12.
7. Bachmann R, Leonard D, Delzenne N, Kartheuser A, Cani PD. Novel insight into the role of microbiota in colorectal surgery. Gut. 2017;66:738–49.
8. Manichanh C, Eck A, Varela E, Roca J, et al. Anal gas evacuation and colonic microbiota in patients with flatulence: effect of diet. Gut. 2014;63:401–8.
9. Sarna SK. Colonic motility. From bench side to bed site. San Rafael (CA): Morgan & Claypool Life Sciences; 2010. p. 1–157.
10. Wang XY, Paterson C, Huizinga JD. Cholinergic and nitrergic innervation of ICC-DMP and ICC-IM in the human small intestine. Neurogastroenterol Motil. 2003;15:531–43.
11. Lin AY, Dinning PG, Milne T, Bissett IP, O'Grady G. The "rectosigmoid brake": review of an emerging neuromodulation target for colorectal functional disorders. Clin Exp Pharmacol Physiol. 2017;44:719–28.
12. Bassotti G, Gaburri M. Manometric investigation of high-amplitude propagated contractile activity of the human colon. Am J Physiol. 1988;255:G660–4.
13. Rao SS, Sadeghi P, Beaty J, Kavlock R, Ackerson K. Ambulatory 24-h colonic manometry in healthy humans. Am J Physiol Gastrointest Liver Physiol. 2001;280:G629–39.
14. Bassotti G, Gaburri M, Imbimbo BP, Rossi L, Farroni F, Pelli MA, Morelli A. Colonic mass movements in idiopathic chronic constipation. Gut. 1988;29:1173–9.

15. Narducci F, Bassotti G, Gaburri M, Morelli A. Twenty four hour manometric recording of colonic motor activity in healthy man. Gut. 1987;28:17–25.
16. Dinning PG, Fuentealba SE, Kennedy ML, Lubowski DZ, Cook IJ. Sacral nerve stimulation induces pan-colonic propagating pressure waves and increases defecation frequency in patients with slow-transit constipation. Colorectal Dis. 2007;9:123–32.
17. Fukumoto S, Tatewaki M, Yamada T, Fujimiya M, Mantyh C, Voss M, Eubanks S, Harris M, Pappas TN, Takahashi T. Short-chain fatty acids stimulate colonic transit via intraluminal 5-HT release in rats. Am J Physiol Regul Integr Comp Physiol. 2003;284:R1269–76.
18. Weeks JC, Nelson H, Gelber S, Sargent D, Schroeder G, Clinical Outcomes of Surgical Therapy (COST) Study Group. Short-term quality-of-life outcomes following laparoscopic-assisted colectomy vs open colectomy for colon cancer: a randomized trial. JAMA. 2002;287:321–8.
19. Walter CJ, Collin J, Dumville JC, et al. Enhanced recovery in colorectal resections: a systematic review and meta-analysis. Colorectal Dis. 2009;11:344–53.
20. Greco M, Capretti G, Beretta L, Gemma M, Pecorelli N, Braga M. Enhanced recovery program in colorectal surgery: a meta-analysis of randomized controlled trials. World J Surg. 2014;38:1531–41.
21. Grass F, Schäfer M, Demartines N, Hübner M. Normal diet within two postoperative days-realistic or too ambitious? Nutrients. 2017;9:1336–45.
22. Han-Geurts IJ, Hop WC, Kok NF, Lim A, Brouwer K, Jeekel J. Randomized clinical trial of the impact of early enteral feeding on postoperative ileus and recovery. Br J Surg. 2007;94:555–61.
23. Zhang K, Cheng S, Zhu Q, Han Z. Early versus traditional postoperative oral feeding in patients undergoing elective colorectal surgery: a meta-analysis of safety and efficacy. [Article in Chinese]. Zhonghua Wei Chang Wai Ke Za Zhi. 2017;20:1060–6.
24. Hassan I, Pemberton JH, Young-Fadok TM, You YN, Drelichman ER, Rath-Harvey D, Schleck CD, Larson DR. Ileorectal anastomosis for slow transit constipation: long-term functional and quality of life results. J Gastrointest Surg. 2006;10:1330–6.
25. Ripetti V, Caputo D, Greco S, Alloni R, Coppola R. Is total colectomy the right choice in intractable slow-transit constipation? Surgery. 2006;140:435–a440.
26. Vergara-Fernandez O, Mejía-Ovalle R, Salgado-Nesme N, Rodríguez-Dennen N, Pérez-Aguirre J, Guerrero-Guerrero VH, Sánchez-Robles JC, Valdovinos-Díaz MA. Functional outcomes and quality of life in patients treated with laparoscopic total colectomy for colonic inertia. Surg Today. 2014;44:34–8.
27. Bove A, Bellini M, Battaglia E, Bocchini R, Gambaccini D, Bove V, Pucciani F, Altomare DF, Dodi G, Sciaudone G, Falletto E, Piloni V. Consensus statement AIGO/SICCR diagnosis and treatment of chronic constipation and obstructed defecation (part II: treatment). World J Gastroenterol. 2012;18:4994–5013.
28. Li F, Fu T, Tong W, Zhang A, Li C, Gao Y, Wu JS, Liu B. Effect of different surgical options on curative effect, nutrition, and health status of patients with slow transit constipation. Int J Colorectal Dis. 2014;29:1551–6.
29. Feng X, Su Y, Jiang J, Li N, Ding W, Wang Z, Hu X, Zhu W, Li J. Changes in fecal and colonic mucosal microbiota of patients with refractory constipation after a subtotal colectomy. Am Surg. 2015;81:198–206.
30. Reshef A, Gurland B, Zutshi M, Kiran RP, Hull T. Colectomy with ileorectal anastomosis has a worse 30-day outcome when performed for colonic inertia than for a neoplastic indication. Colorectal Dis. 2013;15:481–6.
31. Lillehei RC, Wangensteen OH. Bowel function after colectomy for cancer, polyps, and diverticulitis. JAMA. 1955;159:163–70.
32. Sarli L, Costi R, Sarli D, Roncoroni L. Pilot study of subtotal colectomy with antiperistaltic cecoproctostomy for the treatment of chronic slow-transit constipation. Dis Colon Rectum. 2001;44:1514–20.
33. Costi R, Roncoroni L, Violi V, Sarli L. Subtotal colectomy with antiperistaltic cecoproctostomy for slow-transit constipation: concerning the paper: Jiang CQ, Qian Q, Liu ZS, Bangoura G, Zheng KY, Wu YH. Subtotal colectomy with antiperistaltic cecoproctostomy for selected patients with slow-transit constipation-from Chinese report. Int J Colorectal Dis 2008; 23: 1251–1256. Int J Colorectal Dis. 2009;24:1117–8.

34. Yang D, He L, Su TR, Chen Y, Wang Q. Outcomes of laparoscopic subtotal colectomy with cecorectal anastomosis for slow-transit constipation: a single center retrospective study. Acta Chir Belg. 2018;27:1–5.

35. Malone PS, Ransley PG, Kiely EM. Preliminary report: The antegrade continence enema. Lancet. 1990;336:1217–8.

36. Hill J, Stott S, MacLennan I. Antegrade enemas for the treatment of severe idiopathic constipation. Br J Surg. 1994;81:1490–1.

37. Marshall J, Hutson JM, Anticich N, Stanton MP. Antegrade continence enemas in the treatment of slow-transit constipation. J Pediatr Surg. 2001;36:1227–30.

38. Altomare DF, Rinaldi M, Rubini D, Rubini G, Portincasa P, Vacca M, Artor NA, Romano G, Memeo V. Long-term functional assessment of antegrade colonic enema for combined incontinence and constipation using a modified Marsh and Kiff technique. Dis Colon Rectum. 2007;50:1023–31.

39. Phillips SF, Quigley EM, Kumar D, Kamath PS. Motility of the ileocolonic junction. Gut. 1988;29:390–406.

40. Dinning PG, Bampton PA, Kennedy ML, Kajimoto T, Lubowski DZ, de Carle DJ, Cook IJ. Basal pressure patterns and reflexive motor responses in the human ileocolonic junction. Am J Physiol. 1999;276:G331–40.

41. Miller LS, Vegesna AK, Sampath AM, Prabhu S, Kotapati SK, Makipou K. Ileocecal valve dysfunction in small intestinal bacterial overgrowth: a pilot study. World J Gastroenterol. 2012;18:6801–8.

42. Van Citters GW, Lin HC. The ileal brake: a fifteen-year progress report. Curr Gastroenterol Rep. 1999;1:404–9.

43. Pucciani F, Bologna A, Cianchi F, Cortesini C. Anorectal physiology following sphincter-saving operations for rectal cancer. Dig Surg. 1993;10:33–8.

44. Bassotti G, de Roberto G, Chistolini F, Morelli A, Pucciani F. Case report: colonic manometry reveals abnormal propulsive behaviour after anterior resection of the rectum. Dig Liver Dis. 2005;37:124–8.

45. Williamson MER, Lewis WG, Holdsworth PJ. Decrease in the anorectal pressure gradient after low anterior resection of the rectum. Dis Colon Rectum. 1994;37:1228–31.

46. Pucciani F. A review on functional results of sphincter-saving surgery for rectal cancer: the anterior resection syndrome. Updates Surg. 2013;65:257–63.

47. De la Fuente SG, Mantyh CR. Reconstruction techniques after proctectomy: what's the best? Clin Colon Rectal Surg. 2007;20:221–30.

48. Farouk R, Duthie GS, Lee PW, Monson JR. Endosonographic evidence of injury to the internal anal sphincter after low anterior resection: long term follow-up. Dis Colon Rectum. 1998;41:888–91.

49. Brown SR, Seow-Choen F. Preservation of rectal function after low anterior resection with formation of a neorectum. Semin Surg Oncol. 2000;19:376–85.

50. Stelzner S, Böttner M, Kupsch J, Kneist W, Quirke P, West NP, Witzigmann H, Wedel T. Internal anal sphincter nerves—a macroanatomical and microscopic description of the extrinsic autonomic nerve supply of the internal anal sphincter. Colorectal Dis. 2018;20:07–016.

51. Pescatori M. Myoelectric and motor activity of the terminal ileum after pelvic pouch for ulcerative colitis. Dis Colon Rectum. 1985;28:246–53.

52. Lim M, Sagar P, Finan P, Burke D, Schuster H. Dysbiosis and pouchitis. Br J Surg. 2006;93:1325–34.

53. Mimura T, Rizzello F, Helwig U, Poggioli G, Schreiber S, Talbot IC, Nicholls RJ, Gionchetti P, Campieri M, Kamm MA. Four-week open-label trial of metronidazole and ciprofloxacin for the treatment of recurrent or refractory pouchitis. Aliment Pharmacol Ther. 2002;16:909–17.

54. Aziz O, Athanasiou T, Fazio VW, Nicholls RJ, Darzi AW, Church J, Phillips RK, Tekkis PP. Meta-analysis of observational studies of ileorectal versus ileal pouch-anal anastomosis for familial adenomatous polyposis. Br J Surg. 2006;93:407–17.

55. Khan M, Jayne D, Saunders R. Comparison of defecatory function after laparoscopic total colectomy and ileorectal anastomosis versus a traditional open approach. Ann R Coll Surg Engl. 2018;100:235–9.

56. Ntoni MH, Bouchard LC, Jacobs JM, et al. Stress management, leukocyte transcriptional changes and breast cancer recurrence in a randomized trial: an exploratory analysis. Psychoneuroendocrinology. 2016;74:269–77.
57. Hurtado CG, Wan F, Housseau F, Sears CL. Roles for interleukin 17 and adaptive immunity in pathogenesis of colorectal cancer. Gastroenterology. 2018;155(6):1706–15.
58. Alexander JL, Scott AJ, Pouncey AL, Marchesi J, Kinross J, Teare J. Colorectal carcinogenesis: an archetype of gut microbiota-host interaction. Ecancermedicalscience. 2018;12:865.
59. Yde J, Larsen HM, Laurberg S, Krogh K, Moeller HB. Chronic diarrhoea following surgery for colon cancer-frequency, causes and treatment options. Int J Colorectal Dis. 2018;33(6):683–94.
60. Hamada T, Liu L, Nowak JA, Mima K, Cao Y, Ng K, Twombly TS, Song M, Jung S, Dou R, Masugi Y, Kosumi K, Shi Y, da Silva A, Gu M, Li W, Keum N, Wu K, Nosho K, Inamura K, Meyerhardt JA, Nevo D, Wang M, Giannakis M, Chan AT, Giovannucci EL, Fuchs CS, Nishihara R, Zhang X, Ogino S. Vitamin D status after colorectal cancer diagnosis and patient survival according to immune response to tumour. Eur J Cancer. 2018;103:98–107.
61. David LA, Maurice CF, Carmody RN, Gootenberg DB, Button JE, Wolfe BE, Ling AV, Devlin AS, Varma Y, Fischbach MA, Biddinger SB, Dutton RJ, Turnbaugh PJ. Diet rapidly and reproducibly alters the human gut microbiome. Nature. 2014;505(7484):559–63.
62. Sundin J, Öhman L, Simrén M. Understanding the gut microbiota in inflammatory and functional gastrointestinal diseases. Psychosom Med. 2017;79(8):857–67.
63. Ma Y, Hu M, Zhou L, Ling S, Li Y, Kong B, Huang P. Dietary fiber intake and risks of proximal and distal colon cancers: a meta-analysis. Medicine (Baltimore). 2018;97(36):e11678.
64. Bennet SMP, Böhn L, Störsrud S, Liljebo T, Collin L, Lindfors P, Törnblom H, Öhman L, Simrén M. Multivariate modelling of faecal bacterial profiles of patients with IBS predicts responsiveness to a diet low in FODMAPs. Gut. 2018;67(5):872–81.
65. Fellows R, Denizot J, Stellato C, Cuomo A, Jain P, Stoyanova E, Balázsi S, Hajnády Z, Liebert A, Kazakevych J, Blackburn H, Corrêa RO, Fachi JL, Sato FT, Ribeiro WR, Ferreira CM, Perée H, Spagnuolo M, Mattiuz R, Matolcsi C, Guedes J, Clark J, Veldhoen M, Bonaldi T, Vinolo MAR, Varga-Weisz P. Microbiota derived short chain fatty acids promote histone crotonylation in the colon through histone deacetylases. Nat Commun. 2018;9:105.
66. Ishino K, Mutoh M, Totsuka Y, Nakagama H. Metabolic syndrome: a novel high-risk state for colorectal cancer. Cancer Lett. 2013;334(1):56–61.
67. Lombardi VRM, Carrera I, Corzo L, Cacabelos R. Role of bioactive lipofishins in prevention of inflammation and colon cancer. Semin Cancer Biol. 2017; https://doi.org/10.1016/j.semcancer.2017.11.012.
68. Murray-Stewart T, Casero RA. Regulation of polyamine metabolism by curcumin for cancer prevention and therapy. Med Sci (Basel). 2017;5(4):E38.
69. O'Keefe SJ. Diet, microorganisms and their metabolites, and colon cancer. Nat Rev Gastroenterol Hepatol. 2016;13(12):691–706.
70. Törnblom H, Simrén M, Abrahamsson H. Gastrointestinal motility and neurogastroenterology. Scand J Gastroenterol. 2015;50(6):685–97.
71. Barnes JL, Zubair M, John K, Poirier MC, Martin FL. Carcinogens and DNA damage. Biochem Soc Trans. 2018;46:1213–24.
72. Ma L, Hu L, Feng X, Wang S. Nitrate and nitrite in health and disease. Aging Dis. 2018;9(5):938–45.
73. Nieminen MT, Salaspuro M. Local acetaldehyde-an essential role in alcohol related upper gastrointestinal tract carcinogenesis. Cancers (Basel). 2018;10(1):E11.
74. Clarke JD, Roderick H, Dashwood RH, Ho E. Multi-targeted prevention of cancer by sulforaphane. Cancer Lett. 2008;269(2):291–304.
75. Eggersdorfer M, Wyss A. Carotenoids in human nutrition and health. Arch Biochem Biophys. 2018;652:18–26.
76. Chu C, Deng J, Man Y, Qu Y. Green tea extracts epigallocatechin-3-gallate for different treatments. Biomed Res Int. 2017;2017:5615647.
77. Lange MM, van der Velde CJ. Faecal and urinary incontinence after multimodality treatment of rectal cancer. PLoS Med. 2008;5(10):1–4.

78. Ridlon JM, Wolf PG, Gaskins HR. Taurocholic acid metabolism by gut microbes and colon cancer. Gut Microbiol. 2016;7(3):201–15.
79. Wang X, Wang J, Rao B, Deng L. Gut flora profiling and fecal metabolite composition of colorectal cancer patients and healthy individuals. Exp Ther Med. 2017;13(6):2848–54.
80. Hooper LV, Littman DR, Macpherson AJ. Interactions between the microbiota and the immune system. Science. 2012;336(6086):1268–73.
81. Bergman KR, Liu SX, Tian R, Kushnir A, Turner JR, Li HL, Chou PM, Weber CR, De Plaen IG. Bifidobacteria stabilize claudins at tight junctions and prevent intestinal barrier dysfunction in mouse necrotizing enterocolitis. Am J Pathol. 2013;182(5):1595–606. https://doi.org/10.1016/j.aj-path.2013.01.013. Bromatec pp. 84–59
82. Bamia C, Lagiou P, Buckland G, et al. Mediterranean diet and colorectal cancer risk: results from a European cohort. Eur J Epidemiol. 2013;28:317–28.
83. Drago S, Asmar R, Di Pierro M, et al. Gliadin, zonulin and gut permeability: effects on celiac and non- celiac intestinal mucosa and intestinal cell lines 1. Scand J Gastroenterol. 2006;41:408–19.
84. Pucciani F. Post-surgical fecal incontinence. Updat Surg. 2018;70(4):477–84.
85. Buckman SA, Heise CP. Nutrition considerations surrounding restorative proctocolectomy. Nutr Clin Pract. 2010;25(3):250–6.
86. Irwin MR, Cole SW. Reciprocal regulation of the neural and innate immune systems. Nat Rev Immunol. 2011;11:625–32.
87. Schieffer KM, Williams ED, Yochum GS, Koltun WA. Review article: the pathogenesis of pouchitis. Aliment Pharmacol Ther. 2016;44(8):817–35.

Nutritional Support After Surgery for Proctologic Diseases

11

Arcangelo Picciariello and Maria Teresa Rotelli

11.1 Haemorrhoid Treatment

11.1.1 Indication

Haemorrhoidal disease is the most common anorectal disorder, affecting 5–10% of the population.

Considering the high but unknown prevalence of haemorrhoids, especially in the age between 45 and 65 years, they represent an important medical and socioeconomic problem.

The true pathophysiology is poorly understood, but the sliding anal canal lining associated to altered collagen and predisposing factors like constipation or chronic diarrhoea is the most accepted theory (Thompson) [1].

During the last decades, different classifications have been proposed, but the most widely accepted is still the anatomical Goligher's grade classification [2]:

- I degree: haemorrhoids project slightly into the lumen of the anal canal with bleeding.
- II degree: haemorrhoids protrude into the anal canal and descend towards the anal orifice, appearing externally during straining, but return spontaneously after defecation.
- III degree: haemorrhoids protrude externally during the defecation and require a manual manoeuvre to reposition into the anus.
- IV degree: haemorrhoids are always outside the anus and are irreducible.

According to the European Guidelines for Haemorrhoids, their treatment depends on the Goligher's grade and associated symptoms, such as bleeding, pain and itch

A. Picciariello (✉) · M. T. Rotelli
Department of Emergency and Organ Transplantation, University of Bari, Bari, Italy

© Springer Nature Switzerland AG 2019
D. F. Altomare, M. T. Rotelli (eds.), *Nutritional Support after Gastrointestinal Surgery*, https://doi.org/10.1007/978-3-030-16554-3_11

(https://www.escp.eu.com/guidelines#haemorrhoids   https://www.escp.eu.com/guidelines/processes-and-methods).

Medical treatment is based on high-fibre diet or fibre supplements and the use of vasoactive flavonoids combination, including diosmin, hidrosmin, hesperidin and rutosides. Although many authors reported good results, the mechanism of these drugs remains unclear [3].

The surgical management of haemorrhoids has changed significantly in the last two decades with the introduction of new surgical devices and procedures.

After surgery for haemorrhoids, patient education is mandatory, including analgesia, softening of the stools, warm sitz baths and information concerning early and late complications of each procedure.

11.1.2 Surgical Techniques

11.1.2.1 Rubber Band Ligation (RBL)
Rubber band ligation is the most common noninvasive office-based procedure for grade II and III haemorrhoids. This technique is easy to perform, effective and quite safe. Even if complications such as vasovagal reaction, bleeding, pelvic sepsis and Fournier's gangrene have been reported, they are very rare and uncommon.

This technique consists of grasping the internal haemorrhoidal cushion in the area above the dentate line checking the absence of sensitivity. Then, using a proctoscope, the cushion is pulled through a metal ring, and a single or double rubber band is applied. Single or multiple ligations can be performed in a single session.

The success rate of this technique for grade II–III haemorrhoids ranges from 69 to 97% with variable recurrence rate (6–18%) [4].

11.1.2.2 Stapler Haemorrhoidopexy (PPH)
Stapler haemorrhoidopexy is mostly performed for grade III haemorrhoids (Fig. 11.1).

Fig. 11.1 III degree haemorrhoids

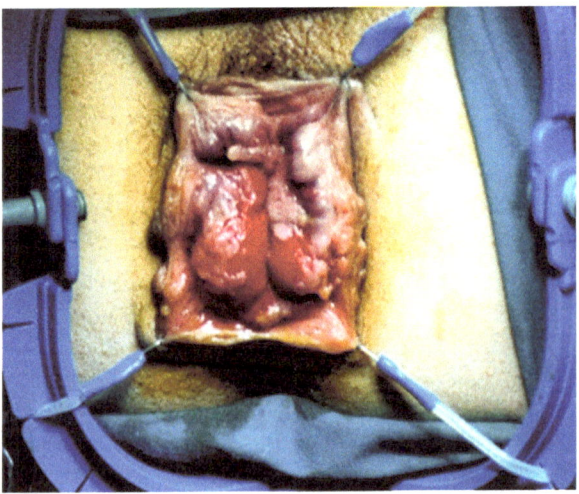

Fig. 11.2 Anal canal after stapled haemorrhoidopexy

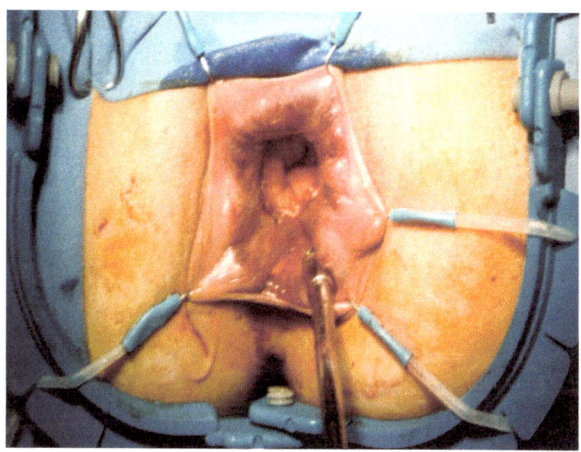

It is defined as the use of a transanal circular stapler to excise a complete circular strip of rectal mucosa approximately 4 cm proximal to the dentate line removing the redundant mucosa (including a part of the muscular layer of the lower rectum) and stapling off the end of the branches of superior haemorrhoidal arteries. In this way, the prolapsed haemorrhoidal tissue is lifted proximal to the dentate line (Fig. 11.2).

According to the Thomson theory, this procedure minimizes postoperative pain and avoid scar, sparing the sensitive anoderm.

On the other hand, rare but severe early complications have been reported such as rectovaginal fistulas, retroperitoneal sepsis, bleeding and faecal urgency. Late complications are faecal incontinence chronic pelvic pain and rectal pocket syndrome.

Probably severe early complications and the poor long-term outcome of this operation are the reasons why in the last years, there is a decrease tendency to perform this operation [5].

11.1.2.3 Doppler and Not Doppler-Guided Haemorrhoidal Artery Ligation with Mucopexy (DGHAL and NDGHAL with Mucopexy)

Both Doppler and not Doppler-guided haemorrhoidal artery ligations with mucopexy represent non-excisional treatment for haemorrhoids. Best indications of these techniques are II–III degrees of haemorrhoids.

These procedures are based on the principle that ligation of haemorrhoidal artery provides a significant reduction of the arterial flow of the haemorrhoidal cushions.

Furthermore, the mucopexy allows to reallocate cushions in the original site by the plication of the redundant mucosa and submucosa.

Using a proctoscope with or without the Doppler probe, all six main branches of haemorrhoidal arteries (hours 1, 3, 5, 7, 9 and 11 of the anal clock) are ligated with a Z-stitch among 1 cm over the dentate line. Then a mucopexy is performed by a longitudinal runny suture from proximal to distal margin including the redundant mucosa and submucosa of each cushion.

Currently there is no significant evidence of a better outcome due to the Doppler probe even if most of studies report results with the Doppler guide [6].

Since there is no resection or excision of cushions, complications after DGHAL and not-DGHAL are very rare and include bleeding and tenesmus with anal discomfort. Furthermore, postoperative pain is well controlled.

On the other hand, recurrence rate is higher compared with procedures involving the excision of haemorrhoidal cushions [7].

11.1.2.4 Haemorrhoidectomy

Classic Haemorrhoidectomy (Milligan-Morgan)

Milligan-Morgan operation, described in 1937, is the traditional open surgical procedure performed for grade II–III and IV haemorrhoids.

With patient in lithotomy or jack-knife position, a circular anal dilatator allows to identify prolapsing haemorrhoidal cushions.

Preserving a mucocutaneous bridge between cushions, a dissection of each cushion and ligation by absorbable suture of haemorrhoidal pedicles is performed to avoid intraoperative bleeding.

Then the excision of cushions is carried out using scissors, electrocautery or, preferably, other devices such as LigaSure and the Harmonic scalpel (Fig. 11.3) that allows to reduce time of operation and postoperative pain and a faster return to normal life.

The most important early complication of Milligan-Morgan operation is postoperative bleeding followed by infection and faecal impaction. Late complications are rare and can be persisting bleeding, anal incontinence and anal strictures.

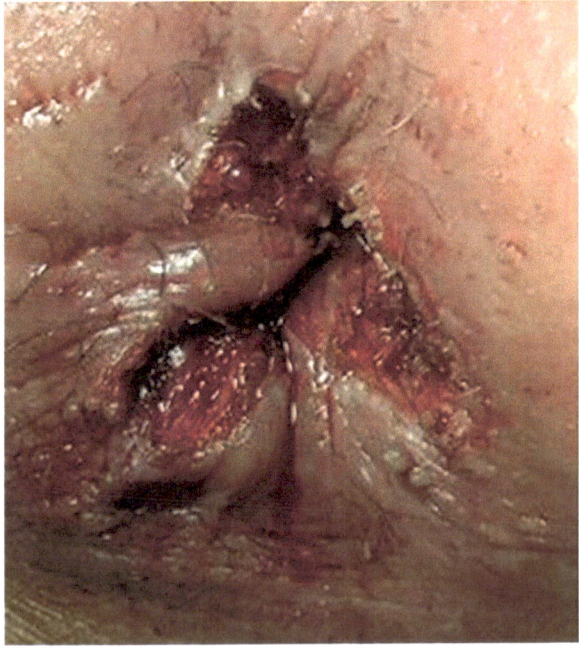

Fig. 11.3 Anus following open haemorrhoidectomy

Even if this procedure causes postoperative pain and a slower return to normal life, it remains the only totally radical operation with a negligible percentage of recurrence [8].

Closed Haemorrhoidectomy (Ferguson)

Ferguson's procedure, described in 1959, consists of closing of mucosa, anoderm and skin with running absorbable sutures, after the ligation and excision of the haemorrhoidal cushions.

Recent systemic reviews and meta-analysis have showed that closed haemorrhoidectomy compared with the open procedure allows a reduced postoperative pain and bleeding and a faster wound healing [9].

11.1.3 Pathophysiologic Implications

All the operations on the anus, including surgery for haemorrhoids but also for anal fissure and fistulas, implicate anal pain of various severity and endurance. This pain is exacerbated by the defecation, particularly when hard stools must be evacuated. The passage of hard stools and strain at defecation can also cause bleeding in case of open haemorrhoidectomy or external haemorrhoidal thrombosis in patient treated with other minimally invasive procedures.

Therefore, any dietetic recommendation should be aimed to have a regular daily bowel movement with soft and lubricated stools and avoid prolonged straining during defecation.

11.2   Nutritional Support After Surgery for Proctologic Diseases

Therapeutic treatment of proctologic diseases always requires dietary intervention depending on symptoms' grade and severity.

Lifestyle modification is helpful to prevent or reduce disease-generated discomfort. Regular physical activity and correct intake of water and dietary fibres (that usually means increase of both) are the first-line approaches to prevent and manage anal problems and constipation, which are always the cause and/or consequence of proctologic diseases.

However, although constipation has been traditionally indicated as one of the main risk factors, some authors found a positive correlation between haemorrhoids and diarrhoea, suggesting a more complex mechanism that justifies a diversified nutritional approach [10].

The postoperative treatment of haemorrhoids, fissure and fistulas benefit of a nutritional intervention aimed to:

- Minimize straining during defecation
- Reduce canal irritation due to frequent stool loose
- Minimize inflammation and promote wound healing
- Avoid disease recurrence

In this perspective, restoring the gastrointestinal health and improving bowel habits represent primary nutritional targets to follow.

Dry and hard stools are common causes of anal discomfort. Patients suffering for hard and dry stools commonly follow a diet lacking in fibre-rich food (whole grains, vegetables, legumes, fruit) [1, 11]. However, such faeces result from the simultaneous deficiency of fluids and fibre, and, accordingly, increasing the fibre intake does not per se result in symptoms relief if is not accompanied by an adequate water intake. Daily fluid intake should be no less than 2000 mL/day. Drinking daily a bicarbonate-sulphate-calcium-enriched mineral water at awakening (i.e. 400 mL/day, Acquasanta Terme di Chianciano, bain-marie warmed at 30–35 °C, for 15 days) accelerates the oro-caecal transit time [12] and exerts positive effects on the gastrointestinal tract, promoting the production of well-formed stools.

Once a correct hydration has been accomplished, a correct fibre intake is essential. Anderson et al. stated that "dietary fibers affect the entire gastrointestinal tract from the mouth to the anus" [13] since it increases the bowel movement, promotes regularity and benefits many gastrointestinal disorders.

Of note, a correct fibres intake could be of benefit also in patients with loose stools since fibres work by absorbing fluid, thereby reducing fluid-related diarrhoea [11]. Soluble and insoluble fibres have different impacts on constipation compared with diarrhoea. Insoluble fibres (whole grain, nuts, green leafy vegetables) act by shortening the food transit time in the gut and should be avoided in patients with diarrhoea, whereas a correct intake of soluble fibres (oats, peas, apples, citrus fruits, carrots, barley, rice) is suggested. Unfortunately, many patients with loose stool look at fibres exclusively as a constipation remedy, becoming incompliant about their introduction in diet.

Dietary fibres may influence the gut microbiota. Soluble fibres such as inulin (naturally present in chicory roots, artichoke, yacon, asparagus, leek, onion, banana, wheat and garlic) have been shown to support the proliferation of "good strains colon bacteria" (*Bifidobacteria* and *Lactobacilli*), thus acting as prebiotic [14, 15].

The current recommendations for the fibres intake prescribe that 14 g/1000 kcal should be introduced daily; this means approximately 28 g/day for adult women and 36 g/day for adult men, including soluble and insoluble fibres. Increasing fibres intake, when requested, must be done gradually in order to minimize the most common adverse gastrointestinal effects of fibres (bloating or excessive gas production) (web sites can help to calculate the amount of fibre per serving, e.g. www.national-fibercouncil.org) [13].

Minimizing tissue inflammation and promoting the surgical wound healing are impelling therapeutic targets for patients and clinicians.

Since dietary added sugar promotes inflammation, an anti-inflammatory diet should minimize the intake of sugar and artificial sweeteners but also of processed food, alcohol, preservatives and excess of dairy products. On the other hand, foods containing probiotic microorganism (yogurt or kefir) are usually recommended to improve the bowel habits reducing dysbiosis and inflammation [16]. Fermented milk, containing *Lactobacillus casei* Shirota, has been recently shown to accelerate haemorrhoid recovery in women during puerperium [17]. Several studies report an inverse association between serum inflammatory markers and dietary fibre intake [18]. In preclinical research oat β-glucan consumption was shown to exert a beneficial effect on immunomodulation, collagen deposition and re-epithelialization, favouring the wound healing process [19].

Consumption of blueberries and raspberries, which are both low-level sugar fruits containing flavonoids, helps to inhibit arachidonic acid metabolism resulting in anti-inflammatory and anti-thrombogenic effects. Postoperative bleeding and haemorrhoid recurrence could also be reduced by flavonoids' protective effect on capillary fragility and permeability [20, 21].

Vitamin A- and Vitamin C-rich foods are suggested to be included in the diet of these postsurgical proctologic patients, because of their role in enzymatic processes related to wound healing [22]. Monitoring of Vitamin D levels could be also helpful. The good source of Vitamin A and Vitamin C found in chilli pepper could probably explain the unexpected results on the presurgical haemorrhoidal symptoms reported by Altomare et al., whereas a prospective, randomized, placebo-controlled, double-blind, crossover trial showed an increasing of symptoms in patients suffering of anal fissure [23, 24].

The use of spices such as curcuma and ginger, promising ingredients of novel functional foods, could modulate the postsurgical inflammatory state, due to their antioxidant and anti-inflammatory properties [25]. Finally, the consumption of extra virgin olive oil, used as dressing and source of "good lipids," antioxidants and vitamins (A, E), has been shown to exert a potent anti-inflammatory action and to be thus beneficial in wound healing [26].

To conclude, a postsurgical educational nutritional program, aimed to disease recurrence prevention and fast recovery, should also promote the importance of a varied and balanced diet providing all macro- and micronutrient intakes [27]. Unfortunately, most of the patients tend to limit food variety because of their vulnerable feelings and fear of pain during defecation, but this kind of mood is strongly counteractive only resulting in further constipation triggering and wound healing delaying time [28].

Freedom of the bowels is the most precious, perhaps even the most essential, of all freedoms – one without which little can be accomplished–

Émile Gautier, 1909

References

1. Lohsiriwat V. Hemorrhoids: from basic pathophysiology to clinical management. World J Gastroenterol. 2012;18(17):2009–17.
2. Gerjy R, Lindhoff-Larson A, Nystrom PO. Grade of prolapse and symptoms of haemorrhoids are poorly correlated: result of a classification algorithm in 270 patients. Color Dis. 2008;10(7):694–700.
3. Alonso-Coello P, Zhou Q, Martinez-Zapata MJ, Mills E, Heels-Ansdell D, Johanson JF, et al. Meta-analysis of flavonoids for the treatment of haemorrhoids. Br J Surg. 2006;93(8):909–20.
4. Iyer VS, Shrier I, Gordon PH. Long-term outcome of rubber band ligation for symptomatic primary and recurrent internal hemorrhoids. Dis Colon Rectum. 2004;47(8):1364–70.
5. Altomare DF, Picciariello A, Pecorella G, Milito G, Naldini G, Amato A, et al. Surgical management of haemorrhoids: an Italian survey of over 32 000 patients over 17 years. Color Dis. 2018;

6. Ratto C, Campenni P, Papeo F, Donisi L, Litta F, Parello A. Correction to: transanal hemor-rhoidal dearterialization (THD) for hemorrhoidal disease: a single-center study on 1000 consecutive cases and a review of the literature. Tech Coloproctol. 2018;22(3):253.

7. Avital S, Itah R, Skornick Y, Greenberg R. Outcome of stapled hemorrhoidopexy versus doppler-guided hemorrhoidal artery ligation for grade III hemorrhoids. Tech Coloproctol. 2011;15(3):267–71.

8. Moult HP, Aubert M, De Parades V. Classical treatment of hemorrhoids. J Visc Surg. 2015;152(2 Suppl):S3–9.

9. Bhatti MI, Sajid MS, Baig MK. Milligan-Morgan (open) versus Ferguson haemorrhoidectomy (closed): a systematic review and meta-analysis of published randomized, controlled trials. World J Surg. 2016;40(6):1509–19.

10. Johanson JF, Sonnenberg A. Constipation is not a risk factor for hemorrhoids: a case-control study of potential etiological agents. Am J Gastroenterol. 1994;89(11):1981–6.

11. Rakinic J, Poola VP. Hemorrhoids and fistulas: new solutions to old problems. Curr Probl Surg. 2014;51(3):98–137.

12. Gasbarrini G, Candelli M, Graziosetto RG, Coccheri S, Di Iorio F, Nappi G. Evaluation of thermal water in patients with functional dyspepsia and irritable bowel syndrome accompanying constipation. World J Gastroenterol. 2006;12(16):2556–62.

13. Anderson JW, Baird P, Davis RH Jr, Ferreri S, Knudtson M, Koraym A, et al. Health benefits of dietary fiber. Nutr Rev. 2009;67(4):188–205.

14. Roberfroid MB. Introducing inulin-type fructans. Br J Nutr. 2005;93(Suppl 1):S13–25.

15. Shoaib M, Shehzad A, Omar M, Rakha A, Raza H, Sharif HR, et al. Inulin: properties, health benefits and food applications. Carbohydr Polym. 2016;147:444–54.

16. Prado MR, Blandon LM, Vandenberghe LP, Rodrigues C, Castro GR, Thomaz-Soccol V, et al. Milk kefir: composition, microbial cultures, biological activities, and related products. Front Microbiol. 2015;6:1177.

17. Sakai T, Kubota H, Gawad A, Gheyle L, Ramael S, Oishi K. Effect of fermented milk containing lactobacillus casei strain Shirota on constipation-related symptoms and haemorrhoids in women during puerperium. Benef Microbes. 2015;6(3):253–62.

18. Wannamethee SG, Whincup PH, Thomas MC, Sattar N. Associations between dietary fiber and inflammation, hepatic function, and risk of type 2 diabetes in older men: potential mechanisms for the benefits of fiber on diabetes risk. Diabetes Care. 2009;32(10):1823–5.

19. Sang S, Chu Y. Whole grain oats, more than just a fiber: role of unique phytochemicals. Mol Nutr Food Res. 2017;61(7) https://doi.org/10.1002/mnfr.201600715.

20. Filingeri V, Buonomo O, Sforza D. Use of flavonoids for the treatment of symptoms after hemorrhoidectomy with radiofrequency scalpel. Eur Rev Med Pharmacol Sci. 2014;18(5):612–6.

21. Panche AN, Diwan AD, Chandra SR. Flavonoids: an overview. J Nutr Sci. 2016;5:e47.

22. Quain AM, Khardori NM. Nutrition in wound care management: a comprehensive overview. Wounds. 2015;27(12):327–35.

23. Altomare DF, Rinaldi M, La Torre F, Scardigno D, Roveran A, Canuti S, et al. Red hot chili pepper and hemorrhoids: the explosion of a myth: results of a prospective, randomized, placebo-controlled, crossover trial. Dis Colon Rectum. 2006;49(7):1018–23.

24. Gupta PJ. Consumption of red-hot chili pepper increases symptoms in patients with acute anal fissures. A prospective, randomized, placebo-controlled, double blind, crossover trial. Arq Gastroenterol. 2008;45(2):124–7.

25. Xu XY, Meng X, Li S, Gan RY, Li Y, Li HB. Bioactivity, health benefits, and related molecular mechanisms of curcumin: current progress, challenges, and perspectives. Nutrients. 2018;10(10):E1553.

26. Parkinson L, Keast R. Oleocanthal, a phenolic derived from virgin olive oil: a review of the beneficial effects on inflammatory disease. Int J Mol Sci. 2014;15(7):12323–34.

27. Sielezneff I, Antoine K, Lecuyer J, Saisse J, Thirion X, Sarles JC, et al. Is there a correlation between dietary habits and hemorrhoidal disease? Presse Med. 1998;27(11):513–7.

28. Andrews CN, Storr M. The pathophysiology of chronic constipation. Can J Gastroenterol. 2011;25(Suppl B):16B–21B.